Down to Earth

Down to Earth

A Dialogue with Mother Nature on the State of the Environment

Michael E. Rice

Writers Club Press

San Jose New York Lincoln Shanghai

Down to Earth
A Dialogue with Mother Nature on the State of the Environment

All Rights Reserved © 2001 by Michael E. Rice

Writers Club Press
an imprint of iUniverse.com, Inc.

For information address:
iUniverse.com, Inc.
5220 S 16th, Ste. 200
Lincoln, NE 68512
www.iuniverse.com

ISBN: 0-595-18116-3

Printed in the United States of America

Contents

Preface

First of all, let me say that I exist. By that I mean I am a 'real' person. The other characters portrayed in this book, however, are fictional—except for 'Lily', I merely changed her name to protect the semi-innocent—and any resemblance to persons living (including any of their past lives for those who believe in such occurrences) or dead (including all present reincarnations, of course) or to any animals, vegetables, or minerals is purely coincidental.* Also, I apologize to those who may feel that I have singled out any one person or group for disparagement. That was not my intent. I tried to be fair and disparage everyone equally.

Secondly, I feel that it is extremely important that the environmental woes facing our planet are brought to the attention of the greatest number of people. That said, know that the book you are about to read contains many facts that you may find disturbing—not the least of which is the attempted destruction of Mother Earth by one of her species: Homo Sapiens (or, more specifically, a subset of that species, Homo Destructus). And, since some of you may find the subject matter of this book hard to swallow, I have tried to spice it up in such a way as to make it a bit more palatable without altering the facts. At any rate, I hope that you keep an open mind as you read this, that you enjoy yourself, and most of all, that you come away with a better understanding of what is taking place on the planet (regardless of whether it is your permanent home or you are just visiting).

Lastly, I would be remiss if I failed to acknowledge the contributions of a few people who were instrumental in bringing this book to fruition and to offer them my thanks: W. Lisa Houser for her help with research, her many readings of this work in its various forms, the beautiful painting that serves as the cover for this book, and her continued encouragement; to my family for their love and support; to the staff at iUniverse for their many contributions (especially Jon McWilliams & Kerry Mickle); to the friends of Bill W. for their continued reality checks; and last, but by no means least, to all of the heroes who have given their blood, sweat, and tears and, in some cases, their lives to protect and defend Mother Earth and her inhabitants. You have my eternal gratitude, one and all. Now, on with the story.

---

*   I'm saying this under protest at the 'urging' of my attorney (who is acting in league with my family and my psychiatrist). Personally (author says after carefully looking both ways and lowering his voice to a whisper) I believe that *all* of the characters in this book are 'real'.

Introduction

This whole thing started innocently enough with a phone call from Lily, my 'significant other' in today's parlance. Her full name is Lillian Lisa Lilly. She has tried for years to get her family and friends to refer to her by her middle name, but we refuse to comply. After all, would you pass on the chance to introduce someone as Lily Lilly? I didn't think so. I mean, come on. Anyway, it was a Thursday morning and I was at work. Our conversation went something like this:

"This is Mike, can I help you?" I said, answering the telephone.

"Hi, it's me," Lily said. "Can you call me back?" She sounded excited.

"Sure," I said.

"Okay." She hung up. I dialed her number. (No, neither one of us is in the CIA nor are we drug dealers—although it has been suggested by more than one person that perhaps Lily could benefit from some form of medication. But, I won't go into that now. I think that will become quite clear as this story moves along. The telephone hi-jinks were simply a result of having a WATTS line where I work and my long distance calls being free. Lily lives in a small town just outside of my city's local call range.) She answered on the first ring. "You there?" she asked.

"Yes," I answered. You are probably thinking what a question for her to have asked me. Who else would it have been? But, in her defense, she wears one of those hands-free telephone headsets like you see stockbrokers and SWAT team members wearing on television. (She is neither one of those, either. She owns her own business, a school for DUI offenders, and only has to work two weekends per month). She wears the headset because she likes to have her hands free to perform other tasks while she is talking on the telephone. Things like typing on her computer or doing housework. She will call me, tell me something, and then ask me what I think. I will respond and there will follow a long gap of silence (except for the sound of a keyboard clacking or dishes being rinsed off). Then she will mumble something like: "There, that's better" or "Okay, I'm back." Multi-tasking she calls it. I call it annoying. The point is the connection on her headset has a tendency to short out at times, hence the 'You there?'

"Some guy e-mailed me today," she said. "And wants me to do some editing and rewrites on a book he is writing. He read my latest editorial online and thinks that I could do it." Lily has had a few op-ed pieces about environmental issues published at different online Web sites and the local newspaper.

"Well?" she said after a moment of silence.

"Sorry, I was listening for the vacuum."

"What?"

"Nothing," I said. "That's great. What is the book about?"

"Sustainable development."

"Okay, I give. What's sustainable development?" I didn't recall any of her editorials being on that subject.

"I'm not entirely clear on that yet." Explains why it wasn't the subject of any of her editorials.

"So," I said. "If you aren't familiar with the topic of his book, why does he want you to work on it?"

"First of all, he is a science guy and his writing is very technical. He wants me to lighten it up and make it more accessible to the general public. Secondly, I didn't exactly tell him that I was unfamiliar with sustainable development."

"I see. But, again, if you are not familiar with the topic?"

"Well, that's kind of why I called," she said.

"What's kind of why you called? I thought you called to share the news that you had been asked to help write a book."

"I did. But, I also wondered if maybe you could help me?"

"How?" I was starting to miss the sound of her doing dishes.

"He e-mailed the introduction and first chapter of the manuscript to me. I read it over and I'm still not exactly sure what sustainable development is—like I said, his writing is very technical."

"And?"

"I was hoping that maybe you could read it and help me decipher it."

If that was all, no big deal, seemed simple enough. At the time I was writing screenplays and short stories. I was not much interested in environmental issues. My day job is at a factory that produces automotive parts. I run a computer tracking system for a contractor of the factory. I was one of only two employees of the contractor working in the state. My employer's home office was in a neighboring state, 250 miles away. I worked in conjunction with one of the factory employees, a union member, and could only work when he worked—part of the union contract. As a result of these circumstances, I had a lot of free time at work. I also had Internet access that no one I worked with knew about. I was told when I started the job that the phone system that was in use in the plant couldn't support Internet activity. This didn't make any sense to me, so one day I brought in my AOL disk from home and loaded it on my computer. After disabling the LAN software, it fired right up. I figured that the story about the phone system not being able to support Internet activity must have been started by plant management to keep the employees from surfing the Web all day. I saw no purpose in exposing the falsity of the rumor to the plant employees. I was sure that would just lead to abuse of the system and the possibility that the phone system would be changed to make the story true. But, I had made the mistake of telling Lily.

"Yeah, sure, I'll look it over," I said.

"Great. And, maybe do some research for me, too?" I thought *'In a perfect world this is where the headset would short out'*. "Michael, are you still there?"

"Yes, I'm here."

"So, will you help me with the research?"

"Why do you need my help with research? I thought you were just going to be rewriting his material."

"I am. But, I also have an idea for a book of my own. And, with rewriting his book and doing my normal stuff, I won't have much time for research."

"What is your idea for a book?" I asked.

"The addiction of consumerism and how that ties in with the environmental destruction of the planet. I'm going to base it on the characters in the Wizard of Oz." I told you the issue of her needing medication would become more clear.

"So, you plan on writing two books at the same time?"

"Yes. I don't think it will be that difficult since they are both dealing with the environment in some respect." *Great, more multi-tasking. Just what she needs.* "So, will you help me?"

"Sure, why not." Like I said, I had a lot of free time at work.

"Great. I'll attach his stuff to an e-mail and send it right off to you. Look it over and call me later, okay?"

"Okay."

I read the material that Lily had e-mailed, which consisted of a lot of startling and very disturbing facts on the degradation of the environment. I proceeded to have a spiritual meltdown over what part I may have played in it. I had known there were some problems besides what I had read about in Lily's editorials. I had also read some about global warming, the hole in the ozone layer, and certain species being on the endangered list, I wasn't totally clueless, but I had no idea that things were as bad as they were. What I did know was that if I wanted to heal this hole in my soul I had to help Lily with her projects.

I did some preliminary research on the Web while I was still at work. I can't stay online for long periods of time. It's hard to justify a busy signal to the home office when they are generally the only people I have reason to talk to on the telephone. I like to keep my Internet time free to use for e-mailing friends and family, researching stocks, and downloading cool screensavers. You know, important stuff. Plus, I actually have to work occasionally. Go figure. Anyway, after work I headed over to the library. I like this method of research better anyway. I prefer the feel of the books in my hands and the intimacy of a library or a good book store.

The next afternoon, I followed the same procedure and that's when things began to get weird…really weird. It was the middle of a Friday afternoon—I get off at 2:30 p.m.—and the library was not very busy. I was seated in the same place as the day before—at a table I had found tucked away upstairs in a secluded little nook behind the stacks holding the reference books. No windows, any foot traffic, any distractions. Research heaven. There was a stack of books at my elbow, one open in front of me from which I was gleaning some data.

I sensed someone hovering at my shoulder and I looked behind me to find a woman standing there peering down at my notes. I was a bit startled, but recovered quickly and flashed her a brief smile. I didn't grant her my full attention, she appeared harmless, and at first glance only registered a woman of 40 or so in what appeared to be thrift store clothes and carrying a huge macramé purse. She looked like she was going for the 1960's hippie look. I figured she must be into the retro thing. She didn't make any move to leave.

"Could I help you with something?" I said.

"No," she said. "But I can help you. With your permission of course."

"Really, how's that?" I asked. *Why not play along. I'm ready for a break.*

"Well, my name is Mona Tempus." She offered me her hand. I stood up and extended my hand in return. "People of your world know me as Mother Nature." (Like I said, things got weird.)

She grasped my hand before I could pull it back. We shook hands for a few seconds, during which time I stood mute; there was no way I was telling her my name after her revelation. I made a quick assessment of what she had said to me. My assessment was that she must be on a field trip from the local psychiatric hospital and had slipped away from her group. As I pulled my hand away, she walked around the table to stand opposite me. I used this time to look at her more closely. She was rather tall and lithe and the grace with which she moved gave me pause to reconsider. I looked into her eyes for the first time and was immediately captivated by how startlingly blue they were, like an afternoon sky in late spring, and by the depth of intelligence and the hint of power that they conveyed.

I broke away from her gaze long enough to take notice of the rest of her—which was an odd experience in itself. I couldn't get a steady read on her. By that I mean her appearance changed as she moved. At first she had blonde hair and a creamy white complexion to go along with the blue eyes mentioned earlier; then she turned her head slightly and her hair seemed to get darker and longer, her skin color deepened to a rich tan, her eyes faded to black and she appeared to be Native American; from another angle, her dark hair shortened again, hers eyes took on a slanted shape and became the color of jade, and she seemed Asian; finally, the dark hair curled tightly, her skin tone became a deep, rich chocolate color, and she appeared to be of African descent. But, in any of these guises that she seemed to effortlessly morph in and out of, she was breathtaking.

"May I sit down?" she asked. Her appearance had returned to its original blonde, blue eyed state. *Did that just happen? Am I seeing things?*

I absently nodded my head 'yes' and plopped down onto my chair as she softly settled onto hers. She set her bag on the floor beside her chair. She crossed her legs and folded her hands into her lap, the picture of patience and calm. We sat and looked at each other for a moment. I began to wonder what kind of drugs they had given her at the home to allow her to reach such a level of serenity. Then I remembered what she had said. "So, you're Mother Nature, huh?"

"I'm sure that seems a little strange to you," she began.

"You think?"

"But, I assure you it's true," she finished.

"Oh, you assure me. Well, then, it must be true." I was feeling angry. I felt that my intelligence was being insulted. "So, what brings you to the library, Mother Nature? You making sure their periodical section has a copy of 'The Whole Earth Catalog'?"

"This is going to be harder than I thought," she said after emitting a small laugh.

"What do you mean by 'this'?" I asked.

"Convincing you of who I am."

"Lady, I can't tell you what a colossal understatement that is."

"Michael, why don't you just give me a minute and hear me out?"

"There is nothing you could possibly say...hey, how did you know my name?"

"Easy, I'm connected," she replied.

"You're in the Mafia?" A myriad of thoughts went through my head at this point: *When did the Mafia get into the mental health field? Must be something to do with the pharmaceutical end of it. Probably got the idea from 'The Sopranos'. But, most importantly, did I do something to upset them?*

She smiled patiently, kind of like she was the one tolerating the mentally challenged. "No, nothing like that," she said. "I'm connected with the source of 'all'. As are you and every other living thing, but my connection is much stronger at present." *So that's it. She isn't nuts, she's just one of those stoned New Age types. That explains the hippie clothes. She probably drove a Volkswagen van here straight from the Organic Grocery...and smoked a joint in the parking lot. What was next, a lecture on the healing power of crystals? A little aromatherapy? Maybe head back to the commune for a nice bowl of frog entrails and eye of newt to ward off evil spirits?* "No, I'm not on drugs," she said, her indulgent smile still in place. "Nor am I some over-aged hippie. And, I wouldn't dream of harming any animals for any reason, especially not to perform some superstitious ritual."

"All of that showed on my face?" I was amazed by her perception.

"Not all. I intuited the rest. I told you, I'm connected."

"What, that means you can read my thoughts?"

"Pretty much," she said. (Have I mentioned that it got really weird?)

"Okay, what am I thinking now?"

" 'If she gets this right, I'll sh…' Although I'm not going to use your exact wording," she said.

"Oh, wow, that's right. Do it again."

" 'I don't care if she is a freak, this is too cool.' "

"Hee hee. This *is* too cool. Sorry about the 'freak' thing."

"No problem. Now can we quit playing parlor games?" She was clearly getting bored with all of this.

"You have to tell me how you do that." I was jazzed to know the trick. I was already thinking of a hundred ways to get rich behind it.

"I told you already it's not a trick."

"You're connected, yeah, yeah. So, how do I get connected like that?"

"One way would be to do what I'm going to ask you to do. After that you won't care so much about getting rich." *Damn, she's good at this. I'm going to have to watch what I think.* This got another smile from her. "Are you interested or not?" she said.

"Of course I'm interested. How do you do it?"

"Not in my ability to read you, in my offer of help."

"If I accept your help will you teach me to do that?" Help with what exactly I didn't know, but I didn't much care at this point. I had dollar signs dancing in my head.

"It's not something I can teach you. It just happens as a result of going through the process," she said.

"The process. Wonderful." *I knew there had to be a catch.* "What, you want me to go to some sort of touchy-feely environmental encounter group? Is that it?"

"No, but a 12-Step group for paranoia wouldn't be a bad idea."

"Cute," I said. "So what process?"

"The process of becoming aware of what's happening in your world," she answered. "Acceptance of your being connected with the 'all'. Becoming a part of the solution to your world's problems. Things like that."

"Gee, is that all?"

"Pretty much."

"Pretty much, huh? Okay, I'll play along for a while. How can you help me?" *May as well get this out in the open.*

She uncrossed her legs, leaned forward, and rested her forearms on the table. "I can save you the time of doing all of this tedious research and give you all the facts and information you need. And, I can help you disseminate this information in such a way that you will be able to reach the largest possible audience and achieve maximum results."

"You sound like an Anthony Robbins tape." She gave me the smile again. "So, I take it that your 'connection' is in the publishing business?"

"That would be part of the 'all', yes. But, I'm not 'connected' with them in the way you are referring."

"Bummer." *May as well speak her native tongue.*

"It would be much easier," she said. "If you would get past the idea that I am some drug-addled, over-aged hippie who just stumbled across you on my way to the next Grateful Dead concert. This planet is suffering severe environmental damage and the public needs to be made aware of the reality of that before it's too late."

She began to finger a medallion that hung from a silver chain around her neck. She looked distraught. *The kook has a good heart, God love her. I suppose it wouldn't hurt to let her help me. What do I care if she thinks she is Mother Nature? Hell, when I used to drink and do drugs, I thought I was Superman. Plus, I do want to learn that whole intuiting thing.*

"If that will keep you listening to me for now," she said shaking her head. *And, she definitely has the intuiting thing down.*

"So, you have a lot of knowledge of environmental issues?" I asked.

"To say I only had 'a lot' would be one of those colossal understatements you mentioned." *If it's true, what a time saver she could be. And, Lily can't get too upset about my working with this woman if it helps with her projects, right?*

"How long have you been doing this environmental stuff?" I asked.

"Since the beginning." She seemed a little evasive.

"Really. What organization—Greenpeace, Sierra Club, World Wildlife Federation?"

"I began a little before their time." She stifled a laugh.

"Okay, exactly how long have you been doing this?" I growled.

She gave me a nonchalant shrug. "Oh, about 20 billion years."

I started gathering up my stuff and looking around for her handler. Suddenly I regretted having chosen such an isolated area to do my research. "Well, it's been wonderful chatting with you, but I really do need to get going."

"I'm really surprised at you, Michael," she said.

"You're surprised at me? That's rich."

"Yes. My information says that you are an open-minded individual, at least when it comes to matters of spirituality, social issues, and even what you have derisively referred to as New Age matters."

"You're info? What info? Where did you get info on me?" I had come to the conclusion that her knowing my name was no big deal, but this was something else.

"I told you my information comes from the 'all'," she said.

"Here we go with the 'all', again. Just what exactly is the 'all'? A cult? I don't want anything to do with a cult. I don't even like Kool-Aid. And, why does it have information about me?"

"The 'all' is what you would normally associate with God."

"And the hits just keep on coming," I said.

"Let me just say that as for you or anyone else believing that there is a God, it is irrelevant to what we are discussing," she said. "To make things simple, I use the term God because that is the most universally recognized term in your world for the power you attribute everything to. We can call the power anything you like, just give me a name or a term that you are comfortable with, okay?"

"God is as good as any," I conceded. "Like you say, it is the term most people would recognize."

"Very good," she said. "As for those who prefer Allah or Buddha or The Tao or whatever power, they are free to substitute their preference for the term God."

"Fine. So, you have a direct line to God, do you?" I asked.

"Yes. As do you and every other living thing, as I said before. You just don't access it clearly yet."

"Obviously not, seeing as how I don't have any information about you."

"I'll tell you anything you want to know, but you could access it yourself if you worked on it more."

"The only thing I want to know about you," I spat out, "is where your keeper is hiding."

"That was awfully harsh, Michael." She seemed genuinely hurt.

I felt like a heel, but I was angry. I felt violated by all this intuiting she was doing and by her having information about me. And frankly, I was a little bit frightened. "I'm sorry," I said. "But this is getting a little too X-Files even for me."

"Why are you so skeptical?"

"Why? You ask me why! Gee, I don't know, I mean I run into people claiming to have direct access to God like he's on the Internet and saying they are what was it, 20 billion years old, everyday. Why would I be skeptical?" *Where's a guy with a straight jacket when you need him? Or, at least one of the librarians.*

"Again, why? You believe in past lives, do you not?"

"Yes, but."

"And, spirit guides, the after life, astral projection, mass consciousness, divine order, the possibility of life on other planets?"

"Yes, but."

"You have looked into Einstein's theories, quantum mechanics, modern physics such as the Superstring theory, and the possibility of time travel. You have chosen a vegan diet because of morality issues, you believe there is a power at work that guides people in the right direction—God for lack of a better term. Yet when faced with the assertion that I am 20 billion years old or that I can get information from this same God whom you have conversations with and meditate about, your mind slams shut like a steel trap on the leg of an innocent animal. Can you explain that to me?" She settled back in her chair and crossed her arms, waiting for my answer.

"Well, for starters, you don't look a day over 10 billion," I said.

"Skeptical and cynical; the reports were right."

"Reports? First it was just some info now it's reports? Just how much do you know about me?" *For a crazy lady she seems to know a lot. Is she an old girlfriend from my wild days that I don't remember?* (Yes, they were that wild). *Oh, wow, do I have a stalker on my hands?*

"Oh, get over yourself. I know only what I need to know to be able to persuade you to be a part of the solution. I don't know any deep, dark, intimate details of your life. A stalker, please." She smirked.

I can only imagine how red my face must have been at that moment. "Then where did you get this information?"

"You're not going to like the answer."

"Not if it's the 'all'."

"That's the only answer I have for you. Accept it, don't accept, it's your choice."

"I can't just accept something like that. I'm going to need some proof. A whole lot of proof."

She shook her head, clearly frustrated. "Okay, maybe I should give you some background on your planet and myself. Then maybe you will be able to crack open your mind a bit and see that I am for real."

"Fair enough." Though I did think it rather ironic of her to be speaking of cracked minds.

"Then how about 'open your mind to all possibilities'? That work better for you?"

Her intuiting had the potential to cause some major annoyance. Especially since I couldn't intuit back.

She smiled that calm smile. "You may want to take notes."

"Whatever." Not the snappiest of comebacks I know, but that's all I had at the moment. I picked up my pen and slid my legal pad in front of me. I thought at the time that it might be best to just humor her. I didn't know if among her other mental health issues that maybe rage was a problem, too. Or, if she had some of her fellow cultists keeping an eye on me. Or if maybe a Mafia enforcer was lurking around. Plus, I didn't know what she was carrying in that big bag of hers. I mean, it was roomy enough to hold a bazooka…okay, so perhaps the paranoia group wasn't such a bad idea after all.

Anyway, that's how I came to meet Mona. I'd like to say that was the start of a beautiful friendship, but it wasn't quite that simple.

CHAPTER 1

GETTING TO KNOW HER

"Like I said before, my name is Mona. M-O-N-A. Short for MOther NAture. My job is to go from solar system to solar system and help get them in shape to support life. I have been doing this work since just after the first universe was created a little over 20 billion years ago—like your 'Star Wars', it was long ago and far, far away."

I looked over the top of my glasses at her. "Yeah, that opened my mind right up."

She shifted in her seat, annoyed. "You may want to let me string together more than three sentences before you pass judgment."

"Sorry. By all means, continue."

"I learned on the job and for a while it was touch and go. I made some mistakes. That first system I worked on was what you would call a real Murphy's Law kind of experience. Pigs with wings, pine trees in the desert, green skies...little things like that. I got better and better at my profession and corrected the mistakes as I moved from job to job."

"Corrected your mistakes?" I interrupted. "You care to explain the platypus?"

She shot me a look I hadn't seen the likes of since I was in the 4th grade and my mom caught me stealing candy money from her purse. I needed a quick defense. "Come on, that was five sentences."

Her smile had turned to ice. "The platypus was just too cute and quirky to not let evolve, so get over it." She paused, daring me to challenge her. I opted to hold my tongue for the moment. Not that I was intimidated or anything. "Anyway, I arrived here around 8 billion years after the Milky Way galaxy was formed by the Big Bang, the Hand of God, random events, any or all of the above—whatever you are comfortable believing—I'll get into that more later." She flicked her hand to signal that it was insignificant to her which of these explanations I chose to believe. "With me was a crew of specialists that I had cultivated over the years. I was still single then. It wasn't easy finding a man who wasn't intimidated by such a strong, intelligent, independent, take-charge woman such as I." *Not to mention that you are a fruitcake with extra nuts.*

"You just can't let go of the insanity thing can you?" *Whoops.*

"Let's just say," I said. "I'm keeping my mind open to all possibilities."

"Such wit."

"Thank you." We exchanged fake smiles.

"I believe that I was about to tell you about my crew," Mona said.

"I believe that you were."

"Spats (the Spaceman) Stevens was in charge of the atmospheric team. They took care of everything above the Earth including the air quality and the ozone layer. On his crew were Dr. Rodney Zeuss, sky specialist, Phoebe Philo, moon specialist, and Lissa Houser, our resident Sun Goddess."

"Those names sound familiar," I commented.

"They should if you are at all familiar with your various mythologies. We were the inspiration for good many of your ancient deities."

"I get it. Like your Dr. Zeuss, the sky specialist is the Greek God Zeus, god of the skies. And, Lissa is the name of an African Sun Goddess."

"Exactly."

"Cool," I said. *But weird.*

That earned me her patient smile again. Which had the potential to become as annoying as the 'intuiting'.

"May I continue?" she queried.

"By all means."

"Good. Professor Demi More-Terrafirma was our geologist, handling everything to do with the land: soil, rocks, sands, mountains, so on and so forth. A very earthy woman is she and endowed with good strong genes. And, a distant relative of one of your modern day movie stars."

"Really?" I had a thing for the ex-Mrs. Willis.

She raised her eyebrows to indicate 'maybe...maybe not'. "Her team included Sydney Isaac Francis, agricultural specialist and his cousins, the lovely and talented Francis twins, Flo and Fawn," she said. "Flo would lay the groundwork for flowers, meadows, lilies of the fields, that little garden in Babylon. She rocked on fruits, vegetables, and grains, too. Fawn helped me with what would be the creatures of your world, her background being in animal husbandry and genetics.

"Dr. Donald (Positive Don) Agualiere oversaw the oceans and everything that flowed into them. His nickname didn't come from having an optimistic nature, either. It came from him believing he was always right when it came to anything to do with water. As in: me—'Don, are you sure that will work?' Don—'I'm positive! Now kindly go away.' The foreman of his crew was Al (Fearless) Fremass, river master. The crew as a collective was known as Donald's Ducks."

My eyebrows shot up.

"It wasn't a pun back then," she said. Smoothing her dress, she forged ahead with her tale. "The final cog in the wheel was Fat Gene Tempus. You know him as Father Time."

"Father Time?" I asked.

"Just go with it, huh?"

"Oh, sure, no problem. First Mother Nature, now Father Time. So is that Time as in t-i-m-e or t-h-y-m-e?" I gave her my most innocent look.

"Must you be a total dweeb?"

I should have known better. That innocent look didn't work when I tried to lie my way out of the stolen change fiasco, either. I rolled my hand in the universal 'go on' gesture, which she did.

"Fat Gene was assigned to the project to deal with the whole time thing and to serve as a free lance troubleshooter. He is the galactic equivalent of a jack-of-all-trades. His motto is 'If I can't fix it, it ain't broke; and, if it ain't broke, don't try to fix it.' By the way, he wants to know who thought up Daylight Savings Time—spring forward, fall back—what's that all about?"

"Hey, I totally agree with Fat Gene on that," I said. "I hate it when December rolls around and it starts getting dark about lunchtime." *God, I'm acting like there really IS a Father Time. Pretty soon I'll be trying to have conversations with Old Man Winter.*

"For goodness sakes, Michael, everyone knows there's not really an Old Man Winter." That infernal intuiting again.

"Oh, excuse me. What was I thinking? As I sit here talking with 'Mother Nature' about her construction crew."

"That's totally different," she said with the utmost confidence.

"How is that different?" *This ought to be interesting.*

"It's obvious, Michael," she said as if talking to a two-year-old.

"Forgive my ignorance. Explain it to me if you would be so kind."

"Okay. What names do you have for summer, spring, and fall?"

I couldn't think of any and I told her so.

"There you have it," she said, a note of triumph in her voice.

"Have what, for Pete's sake?"

"The proof that there is no Old Man Winter."

"Because I can't think of any names for the other seasons? That proves there is no Old Man Winter? Am I missing something here?"

"Yes. Just think about it. If there really were an 'Old Man Winter'," she actually made those obnoxious quotation mark gestures with her hands, "don't you think it follows that there would be a something, something summer and a so and so spring and a funny name fall?" I started to say something, but she barreled ahead, "The fact that there isn't proves that there's not really an 'Old Man Winter', either."

Again with the quotation marks. Between that and the 'intuiting' I was about one 'Have a Nice Day' away from developing homicidal tendencies. Through gritted teeth I was able to get out, "Gee, now that's really logical."

"I thought so. Now where was I?"

It seems like I had more to say, but for the life of me I can't remember what it might be. How did I get on the side of the debate trying to prove that there was an Old Man Winter? How did I get so dazed and confused? Feminine logic? I again waved for her to go on.

"Each crew had complete autonomy when it came to their respective specialties," she continued. "But, overall, we worked as a team. Fat Gene and his crew provided Spats (the Spaceman) with the figures he needed to finalize the spacing of the various celestial bodies. Hence, his nickname—plus, Old Spats is, shall we say, a bit unconventional." *If she considers this Spats character unconventional, then that poor soul must be full tilt bozo.*

"To perform these calculations Fat Gene and Spats (the Spaceman) used an abacus and a really long tape measure," she continued. "With these ultra sophisticated instruments, they managed to align everything in such a way that the earth would have 'the sun in the morning and the moon at night' and everything else that it would need to sustain life. By the way, Fat Gene can't help but notice that today you have at your disposal some very complex and powerful computers—not to mention the combined brainpower of 6 billion people—yet, you have still managed to make the planet resemble the infield at Yankee stadium after a hard-rock concert. I wonder if your Ms. Morrisette finds that ironic?" *For someone her age, she is certainly up on her rock music.*

She rambled on, "While Fat Gene and Spats (the Spaceman) were busy with their chores, I was busy with my other crews preparing for the planting of trees, the sowing of crops, the seeding of flower beds, population of the waters—generally readying the planet for life to thrive.

"After a while, Fat Gene began to find reasons to hang around where I was working more and more often. He said it was because he wanted to monitor the rotation of the planets around the sun and things like that to guarantee that the plants and animals to come would get the sunshine, rain, nutrients, and such that they would need to survive and flourish once evolution took hold. Truth is known, he was really hanging around to put the moves on yours truly. He had been smitten by my charms eons ago, but had been too shy to make his move." She paused here to give me a rueful smile.

"How did you know? Did you read his mind, too?" I asked.

"I must have overlooked the part of your report that mentioned pettiness."

"Just a thought. No pun intended." I tried the innocent look, again. It didn't work, again.

"His shyness around me," she said. "Was a result of old childhood issues triggered by his reaction to my nickname, 'Mother'. Old issues with his mother being the original cause of his eating disorder. Thank goodness he ran into Freud's great to the 18th power grandfather and got all that worked out. Over time—no pun intended."

"I see I'm not the only petty one around here," I mumbled.

"Excuse me?" she said with a bit of an edge.

"Oh, nothing. You were saying?"

She gave me one of those 'you better mind you p's and q's' looks that I used to get from my grade school teachers. I smiled adorably. (At least I like to think it's adorable.) She smiled back. If you could call it a smile. Whatever it was, it was definitely not meant to be adorable.

"Eventually," she continued. "Fat Gene's persistence paid off and I fell in love with him. I believe now that the relationship that formed between us had been preordained—in the stars, as it were."

I rolled my eyes and she smiled playfully. It seemed I had opened a Pandora's Box with respect (or lack thereof) to puns.

"We were destined to use our relationship," she said. "As a model for the interdependent and circular relationships that exist between the air, water, forests, plants, animals, and, finally, humans in your system and that has since been used to modify systems elsewhere. That reminds me, I need to make a note to put *that* policy under review after what I've seen around here lately."

I didn't want to go there, so I kept my mouth shut.

"Four billion years later," Mona said. "Once our tasks were complete, Fat Gene—who now sports a svelte boyish figure as a result of joining me on a strict vegan diet and will proudly be known from now on as 'Lean Gene'—and I were married and left on our honeymoon."

"Whoa. Back up the wagons," I said. "You were in love with this guy for four billion years before you married him! You have a little trouble with commitment maybe?"

"As you know from your studies, time is relative. Let's not make an issue out of it."

"Okay, but what about the commitment thing?" I wasn't about to let it just drop.

"Like I said, let's not make it an issue." She had that certain edge to her voice, again. I ignored it.

"Hit a nerve I see." This was getting to be fun having her on the defensive.

"You really want to get into a discussion about commitment? You?"

"Hey, I thought you weren't privy to any 'deep, dark, intimate details' of my life?" I said with some indignation.

"Your ability—or lack there of—to make a commitment is something I needed to know about," she said with an air of defiance.

"Well, if you know the rumor—and I stress the word rumor— is that I have a slight problem sometimes making a commitment, why did you approach me in the first place?"

"A slight problem! Sometimes!" She barely got that out before the dam broke on her laughter.

"I don't think I have a problem with commitment." I said defensively. *Wasn't I just on offense?*

She was doubled over and had tears running down her cheeks.

"I'm serious," I said.

She waved me off and gasped, "Your longest relationship in the last twenty-five years has been with a pet dog." I started to respond, but she cut me off. "When it comes to women, you haven't had one last longer than a year." *Aha! Obviously your information about me is flawed. It has failed to turn up my current relationship with Lily.* "Your relationship with Lily," she exclaimed. And, started laughing again.

"What's so funny about that? We've been together for over three years."

"Together for over three years!" she said. "You've known each other that long, granted. Out of that time, you have been together for a total of a year…and I'm probably being generous giving you that."

"Okay, so our relationship has been a bit unorthodox by conventional standards. So what? We are still together aren't we?"

"Together again is more like it," she answered, then mumbled to herself, "A bit unorthodox." Then she started laughing uncontrollably again. *Fine, I won't bother to explain. It's obvious that you are unwilling to listen to reason.* "Listen to reason!" She waved me off and gasped, "Quit, please." I thought she might start to hyperventilate any second.

I waited for this latest bout of hysteria to subside, then said, "I'm glad you are finding this so amusing."

"I'm sorry," she said. I didn't hear much conviction. And, the way she was trying not to break into another round of hysterics didn't help much either. "Really. I am sorry," she was able to say with a straight face. Though I detected a slight tic under her right eye from the effort.

"Fine. How about we just move it along here, huh?" I said.

"Good idea. Gosh, where was I?" she asked as she wiped the tears off her cheeks. Good question. I consulted my notes.

"You and Lean Gene were on your long-awaited honey-moon." It felt good to get in a shot of my own.

"Right, thank you." She settled back into her chair, ignoring my shot—which really takes the fun out of it. "After our honeymoon, we went to the Orion Nebula, our next assignment, to take our special talents to that solar system. We left the care of the Milky Way galaxy in the hands of my brother-in-law, Lester Tempus—known to some of you as Silenus. Against my better judgment I might add. He is the walking punch line to brother-in-law jokes."

She rolled her eyes and shook her head, then continued, "Upon our return recently, to see how our seeds had been sown as it were, we found Lester passed out in an alley behind the Planet Hollywood in L.A. and the Earth in a funk. Can't leave you alone for a minute. Look at what kind of shape this place is in now. It looks like you threw the biggest block party in the history of...well, in history. And, you forgot to clean things up before mom and dad got back. Well, we're back! And, we are not happy campers."

Her voice had taken on an ominous tone and her eyes flared a bright red. She looked like Linda Blair in "The Exorcist" right before she starts to projectile vomit the pea soup. *This can't be good.* I started looking around for something to use as a splash shield. Fortunately, the episode only lasted a moment. Her eyes returned to normal and her voice took on its usual calm tone. "We—Lean Gene, Spats (the Spaceman), Demi, Positive Don, myself, and a lot of other hard working folks, and God or whatever you believe in—created a perfectly balanced system that would have provided all the species of the planet with everything you would ever need to survive and flourish. The universe after all is just one large life system. The problem is it's getting cancer due to the actions of modern humans.

"There was an abundance of all the natural resources. All you had to do was show some respect for the Earth, observe its workings, and act accordingly like your ancestors had been doing successfully for three million years. Instead, you got greedy. You decided that you could use the Earth and all that was on it for your personal playground. You thought that you could manipulate the ecosystems to your exclusive advantage and to heck with all the other species. Well, paraphrasing one of your old television commercials, 'Its not nice to fool with Mother Nature'. Needless to say, you are all grounded—literally. No more exploring other planets until you get this one cleaned up."

She paused for a moment to catch her breath. I tried to digest all of this and to make up my mind how to proceed. My first thought was to excuse myself under the guise of going to the restroom, then find a phone to call the 911 lunatic hotline. I didn't know for sure if she was a run-of-the-mill psycho, a stoned New Ager, a hippie suffering an LSD flashback, or some scary combination of all of the above. But, I was fairly certain that the woman needed professional help. Again, I forgot she could 'read' me. "Before you reserve me a room at the ha-ha hotel," she said. "Why don't you take a few moments to digest this, perhaps come up with a question or two that, if answered to your satisfaction, would convince you of the validity of what I have said?"

"Uh, sure," I stammered. We looked at each other for a moment.

"You don't really believe that I pose a physical threat to you?" she asked.

"You're right, I don't. Just give me moment to think," I said. I tried my best to let go of the thought that she was crazy (which made me wonder if I was) and thought about what she had told me so far. I glanced back at my notes and thought some more. I remembered once how I had heard what I still believe to have been the voice of God talking directly to me. I don't feel it is necessary to go into the details of why God was talking to me or what he said (it was years ago and had nothing to do with my experience with Mona—or my prior drug use). The important thing is that I believe it did happen. I thought about the people who had been put in my life and the books I had read and the experiences I had had over the past few years. I also thought about the fact that I had been brought to this experience because I was doing research for a book (or three) on the environment, which was not something I would have done on my own initiative I don't believe. I thought about the things that Mona knew about me and my life; about my attitudes and assets and shortcomings; her powers; the things I had come to believe about life after death, the existence of 'God', and the other things she had mentioned. I thought about it for what seemed like a long time, but was in reality only a few minutes. The thoughts had come fast; one right on the heals of another. Then, a question popped into my mind and I decided to go with it despite my skepticism. "If what you say is true, why did you stay away so long and let the planet get so screwed up?"

"Again, your planet is not the only one in existence that supports life. In fact, yours is merely one of…well, a whole bunch."

"So there is life on other planets?" I asked.

"Of course."

"What kind of life? Have they visited here?" I couldn't believe that I was giving validity to her statements by asking questions like this. I can't explain it; it just felt like the right thing to do.

"In answer to where: many light years away. As for 'them' visiting here, I believe that is quite obvious by my presence."

"So what planet are you from?" I asked.

"Technically, I'm not really from another planet, per se. Lean Gene and I reside with the 'all' when we're not…traveling," she answered.

"Oh, you reside with the 'all'. I see."

"No, you obviously don't. But, you will one day. You will remember where you came from, or more specifically, re-member with who you came from. But, we are getting off track here. You were asking about 'us' letting the planet get so screwed up." She shifted in her seat and toyed with the chain around her neck, again. Exasperated, she continued, "We check up on things as often as our schedule permits, but ultimately it's your responsibility to take care of what you've been given. My goodness, our last visit was only 13,000 years ago. That's the blink of an eye in the whole scheme of things. At that time, the planet was still in wonderful shape. When we popped back in this time, we couldn't believe that so much had happened in such a short time."

"You keep saying 'we' and you mentioned that you and Lean Gene travel together. Is he here with you now?" I asked.

"If by 'here' you mean on this planet, the answer is yes."

"I meant here," I said as I waved my hand around indicating the library.

"No, he's not here at the library. He doesn't like public places. In fact, he doesn't much like you humans at all right now. So, for the most part, he remains at the house in which we are staying. He does his work from there. I think he's becoming an Internet junkie. But, he is monitoring our conversation and will let me know when he has anything to input."

"Monitoring our conversation? You wearing a wire? You have some kind of transmitter in that bag of yours?"

"What? Oh, that's right, you have that X-Files paranoia thing going. No, I'm not wired nor do I have a transmitter in my bag. Once, again, we are connected—big difference. Next question."

"How did you know I had another question? Never mind." I didn't feel like rehashing the 'intuiting' thing again. "Let's say I choose to play along with you for the moment. What exactly am I being asked to do?"

"We want you to write a book about the environmental ills of your planet." She said this like it was the most obvious thing in the world.

"As I am sure you know since you are so connected," I said with a obvious tone of my own, "I'm doing this research for Lily so she can write one (or two) of those type books."

"And, as I'm sure you know," Mona shot back, "the one she is helping with is about Sustainable Development and the other one she is toying with is about consumer addiction and the environment. Two very worthy topics, but too specific for our needs. We wish to cover a wider range of subjects."

"So, you want me to write a book?" I asked.

"Yes," she said, again with the 'obvious' tone.

"Why me?"

"Why not you?"

"I mean, there are a great many other people who are far more knowledgeable about this issue than I am and who would be far more qualified to write a book of this nature." An obvious pun, I know, but I wasn't going to point it out...to her anyway.

"No need to, I got it." I snorted in exasperation. She smiled at her ability to once again turn things around on me. "But, you're right," she said. "There are people who are far more knowledge-able, who could spit out impressive statistics and theories to dazzle the masses, but that's not what is needed here. In fact, that's been done over and over, but unfortunately, with very negligible results. We prefer someone who is just awakening to the problem, who is not too prejudiced by prior experiences having to do with work or school or media exposure or whatever, who is fresh and open minded. Who has realized that something is seriously wrong with the environment and seems to want to do something about it. Someone with the capacity to be courageous and act nobly." *Wow, I guess she does have accurate information about me. Or, she is trying to butter me up.* "But, who is still basically ignorant." *I'm guessing it was the butter me up option.*

"Quit. You'll spoil me with all of this flattery," I said.

She gave me the full watt smile. "And, who has a sense of humor."

"Thanks for that at least."

"We want everyone to be able to read, understand, and relate to this book. We feel that you will be able to get the hard facts and dry statistics that I can supply you with across to your people, but with a lighter touch than say a scientist or a pure researcher."

"Thank you for your confidence," I began.

"You're welcome," she interrupted.

"But, are you sure I'm the right choice?" I continued. "I mean with my commitment issues and all?"

"As sure as we can be at this time," she answered. "After extensive, and oft times, heated debate we decided that your past relationship and commitment issues could be overcome."

"Oh, how gracious of you and…whomever," I said.

"Thank you. Now, are you convinced I am who I say I am?"

"Uh, no," I said.

"Wonderful." She was annoyed again. Goody. (Okay, so she's right about the pettiness; I can live with that. After all, it is balanced nicely by my courage and nobility).

"So you still doubt me?" she asked.

"Yes. I mean my God; this is a lot to get my mind around even if I do believe in past lives and different dimensions and things of that…nature." Sorry, I will try hard to curtail my penchant for puns.

"All right. I'll concede that," Mona said.

"Thank you."

"If you will concede that it's possible what I'm telling you is true."

"There's always a catch with you, isn't there?" She waited for my answer. "Yes, it's possible."

"Thank you."

"But not damned likely."

She ignored that and asked, "Then if you concede it's possible, will you commit to hearing me out?"

"Commit! Me!" She smiled and, again, waited for my answer. "Okay, I'll listen for now," I replied. I figured that even if she was only some Eco-warrior who had suffered a psychotic break, she just may be able to save me a lot of time on research and have answers to questions I didn't even know to ask yet. Plus, I still wanted to learn how to do that 'intuiting' thing. And, this did have the potential for a few laughs—always an important factor when weighing any decision.

"I guess that will just have to do," she said with what I thought was a hint of annoyance. I wondered briefly if her statement was in reference to my minimal verbal commitment or to what she intuited afterwards from my thoughts.

"Yes, it will just have to do," I said.

"Whatever," she said. Yes, she was clearly annoyed. See, this was getting to be fun already.

CHAPTER 2

INTO THE PAST

After promising Mona that I wouldn't call 911 or make a run for it, I excused myself to go to the restroom. I assume she intuited that I was telling her the truth because she didn't follow me. I washed my hands and splashed water on my face. After toweling off, I stared at myself in the mirror for a moment (not something I often do; at least not that often). Briefly I entertained the thought to do just what I had promised Mona I wouldn't do—run—but, I was able to fight the urge. I returned to the table and found Mona peacefully staring off into space. "I'm glad you chose not to run off," she said.

"What and miss the chance to commune with Mother Nature?"

"Wise choice," she said, smiling. "Now, do you have any questions before we move on?"

"Oh, yeah, I definitely have some questions."

"I thought you might," she said.

"Of course you did," I said. She nodded, waiting for me to continue. So, I did. "You said that you and you co-workers get planets in shape to support life, correct?"

"Correct."

"Then, if you only get the planets in shape to support life how did life on Earth begin in the first place?" *I may as well cut to the chase. If she is Mother Nature, she should know this, right?*

"God created the heavens and the earth. Don't you read the Bible?" she replied.

"I'm not a science fiction fan."

"How cynical. Actually, the chapter of Genesis is quite accurate." *So, she is going to use the Bible as a cop out because she doesn't know the answer.*

"You mean to tell me," I asked. "That God created everything in six days then sat down to watch the NFL on Sunday?"

"In a way. Don't you think God could do that?"

"I imagine He could. The question is did He? Couldn't life have just happened by random chance like some believe?"

"That's possible, just not very likely, as I will get into later. For now, let's just assume that there was some power involved in all of this, shall we?"

"We shall," I replied.

"Good. Now, to get back to your question on how life on Earth began and to keep the religiously inclined interested, I want to show you how similar the story of creation from Genesis in the Bible is to the scientific theory of the Big Bang."

"So the Big Bang really happened?" I asked.

"There is a great deal of scientific evidence to support that theory."

"Very diplomatic."

"No sense alienating anyone," she replied. "I feel that it is essential to keeping as many people reading this with an open mind as is possible and by presenting the material in this manner, or at least trying to reconcile the religious with the scientific, we can do just that."

"Fair enough."

"Good. First, let me begin by saying that the prevailing theory is that when the Big Bang occurred, in the first millisecond matter spewed forth at an almost incalculable rate of speed. As the matter expanded, the rate of speed of the expansion of the universe began to decrease and is still decreasing today even though the universe is still expanding and the galaxies are still moving farther and farther apart. Now, keep this in mind while I show you a comparison between the two schools of thought."

"Yes, ma'am," I said.

She leaned over and rummaged through her bag. *This could take a while. My apartment at college was smaller than her bag.* Finally she pulled out a Bible, opened it, and set it on the table in front of her. She motioned for me to scoot my chair around next to hers so as to see the Bible. I did as she bade and looked to where she was pointing, which was at the beginning of the chapter Genesis.

"You will notice," she began. "That according to Genesis God created the universe on the first day and light separates from dark. According to your scientific view, the Big Bang occurred about 16 billion years ago and marks the creation of the universe. During this event, which lasted approximately 8 billion years, light breaks free as electrons bond to atomic nuclei and the galaxies begin to form. You see the similarities?"

"Sure. Except for the Big Bang part of that lasting 8 billion years opposed to Genesis's one day."

"Bear with me."

"Okie dokie."

"Genesis day two says the heavenly firmaments were formed."

"Heavenly firmaments?" I asked.

"The galaxies."

"Oh, okay."

"Your scientists say that during the next 4 billion years is when the Milky Way galaxy was formed. It was during this time that my crew and I arrived here."

"Wait. This is getting more and more bizarre. How come day one of Genesis supposedly lasted 8 billion years and day two only 4 billion years?" I asked.

"As I stated earlier," Mona explained, "the expansion of the universe has been decreasing every since the Big Bang. So each day of Genesis represents a smaller block of, for lack of a better term, scientific time."

"I see. Kind of like dog years, but on a sliding scale."

"Okay." She shook her head as if to clear it. "Day three Genesis says that oceans and dry land appeared. Your scientists say during the corresponding 2 billion dog years," she winked at my smile, "the Earth cooled, water appeared, and the first forms of life almost immediately followed which were bacteria and algae.

"Genesis day four says that the Sun, the Moon, and the stars became visible in the heavens. Scientists say during those one and one half billion years photosynthesis produced an oxygen-rich atmosphere. Genesis day five has the waters alive with animal life and the appearance of the first reptiles and winged animals. The corresponding 500 million years of scientific time postulates that the waters began to swarm with the first multi-cellular life followed by winged insects. And, lastly day six of Genesis supports the appearance of land animals, then mammals, and then humankind. Scientists set this period as lasting from 250 million years ago to approximately 200,000 years ago and state that during this time your third most recent massive extinction event occurred which destroyed over 90% of the life that had evolved to that time. What followed was the repopulating of the land, including the appearance of the first hominids. Those hominids, of course, eventually evolved into…well, you."

"That is really cool," I said turning to her. "It does establish a tie between the religious and scientific communities." Her face was inches from mine and I was again amazed at how beautiful she was. I realized that I was very attracted to her, her lunacy notwithstanding. In fact, her lunacy was probably part of the attraction considering my track record with women. Of course, she sensed this—she probably didn't even need her powers this time.

"Michael, I'm married, remember?"

"I keep forgetting. You don't wear a wedding ring." *Who are you kidding? That's just a put off line.*

Allow me a brief digression: I don't think of myself as handsome. A few women have said they thought I was—my mother and a few girlfriends over the years—so, I don't give that much weight. It's not with false modesty that I say this. It's honestly the way I see myself. Don't get me wrong, I don't believe myself to be hideous. It's not like I frighten children when I walk the streets. I simply see myself as average. I am of average height (5' 9"), average weight (160), salt and pepper hair and beard (more salt these days) both trimmed short, blue eyes. No distinguishing scars or tattoos. I dress decently, but go more for comfort than fashion. I like to think that I have a good sense of humor and have been blessed with reasonable intelligence. As Lily puts it, 'I present well'. I believe that the 'presents well' explains how I have managed to attract a few lovely women over the years. I have been fortunate in that area to my continued amazement. And, in my mind, I've always felt just that—fortunate and amazed—to attract them. I've never seen what they have professed to see in me. Hence, my reaction to Mona's rebuff.

"It has nothing to do with any of that, Michael," Mona said having obviously intuited my inner dialogue. "I am married. I just don't wear a wedding ring. It's not a custom to do so in my culture. In fact, there are many cultures that don't follow that particular custom."

"Really?" I asked.

"Yes, really," she replied. "It would behoove you specifically and your cultures in general to realize that your ways are not the only ways—much less the only right ways."

Embarrassed, again, I said, "Sorry. Point taken." I scooted back around to my side of the table.

"And, even if I weren't married, there is the matter of your relationship with Lily, as it were."

"I was just looking. It wasn't like a made a pass at you or anything. Jeez."

"And, I'm just saying."

"Fine. Can we just move on?"

"Very well," she said from behind a repressed smile. "As you can see by this comparison, religion and science don't have to be mutually exclusive. Science deals with theories and facts, and the Bible, most would agree, teaches with parables. The point is that science and religion, in some cases, may be saying the same thing. Agreeing with one doesn't mean that you have to disagree with the other. It's all a matter of perspective. In the case of the origin of your universe and its life forms, you could agree with the Genesis version and also the Big Bang version, they are essentially saying the same thing. I don't know why those two communities are at odds anyway. After all, many of the great scientists throughout your history have been men of religion or of strong religious beliefs."

"Really? Like whom?" I asked.

"There's Albert the Great, the Father of Geology, Roger Bacon, the Father of Chemistry, Gregor Mendel, the Father of Genetics, Angelo Secchi, the Father of Astrophysics, and Georges Lemaitre, the inventor of the Big Bang theory that we just discussed. All were either priests or monks. Then you have Copernicus, Galileo, Kepler, and Newton who were all devout Christians."

"Wow, I never knew that."

"Even the great Einstein was said to be a deeply religious man."

"That I knew. He said something along the lines of 'science without religion is lame and religion without science is blind'."

"Right. And, he also believed that God didn't play dice with the universe."

"Uh huh," I agreed. I decide to shift gears here. "So, now that we know the origin of life."

"Hold on. There are other theories," she interjected.

"Other theories?" *Great, more theories.*

"Yes, other theories. You mentioned one earlier."

"I did?"

"Yes. The one about random events."

"Oh, yeah. I have read that it is possible that life happened as the result of random chemical reactions that took place over billions of years in some sort of primordial mix."

"Some of your scientists believe that, yes," she said. "Although, as I stated earlier, it's not very likely. In fact, I should point out that the odds of life on your planet appearing as the result of random chance has been figured to be 10 to the 40,000 power to one—that's 10 with 40,000 zeroes after it."

"That's worse odds than Powerball."

"That would make your Powerball seem like a sure thing. A few other facts might help to dispel the random chance theory. I assume from your studies that you are familiar with the fundamental laws of the universe, yes?"

"Like the Law of Gravity, those kinds of laws?"

She nodded 'yes'.

"Familiar in that I've heard of them," I said. "I am no expert by a long shot, though."

"No matter. Since you mentioned the Law of Gravity we will start there. If the properties of the Law of Gravity were larger by just one part in a hundred, thousand, million, million the Big Bang would have collapsed in on itself. If it were smaller by the same infinitesimal amount, the original dust would have just continued to expand, never forming the planets in your universe."

"No Earth?"

"No Earth."

"So, no humans?"

"Not as you are today, no."

"That wouldn't have been a good thing," I said.

"The jury is still out on that point. But, continuing on with randomness, if the nuclear force was a little weaker, the only chemical that would have formed would have been hydrogen. If it were a little stronger, all hydrogen that did form would have been converted to helium and other heavier elements. No hydrogen means no sun and no water."

"Without the sun we wouldn't be thirsty anyway."

"Without the sun, you wouldn't 'be', period. Lastly, if electromagnetism were weaker, atoms would deteriorate at room temperature. If stronger, no chemical compounds would be possible. Like carbon. Since you are carbon-based life forms, no carbon means no life. As you can see, these fundamental Laws are perfect for the formation of your universe and your planet and to support life."

"Wow, what a coincidence."

"Yes, what a coincidence." She gave me a sardonic look.

"I'm just kidding," I said. "I have been told over and over that coincidence is just a difficult way to spell God."

"I've never heard that one before," she said. "May I use it?"

"As far as I'm concerned."

"Thank you."

"Don't mention it."

"There is also the Superstring Theory or what some of your scientists are hailing as the Theory of Everything," she continued. "This theory bridges the gap between Einstein's general theory of relativity which deals with things on a macrocosmic scale—universes and such; and quantum mechanics, which deal with matter on a microcosmic scale—atoms, photons, quarks. It postulates that, at the most basic level, all forms of matter—animal, vegetable, and mineral—and, the forces we just discussed, originate from the same string-like microscopic form of matter. A superstring is thought to be about a million billion billion—that's 10 to the 20th power—times smaller than an atomic nucleus. The only thing that differentiates animals from vegetables from minerals is the infinite ways in which the strings vibrate. Of course, the Superstring Theory doesn't state how the superstring came about. What—or who—made it."

I tried to stifle a yawn, but was unsuccessful. She didn't seem pleased.

"Am I boring you?"

"A little," I said. That caught her off guard.

"Look," I said. "My impression from what you told me about this book is that you don't want it to get bogged down in scientific jargon and streams of statistics. Right?"

"Right. And, your point?"

"My point is that seems to be where you are heading."

"Well, you're the writer. Fix it when you do your rewrite. That is your job after all." She rolled her eyes, which was taking its place right up there on the annoyance scale next to the 'smile' and the intuiting.

"Whoa," I said. "I haven't said I was going to do this yet. I just agreed to hear you out."

"My mistake. But, as long as you are still hearing me out can we get back to the subject matter?"

"Oh, well, forgive me for the interruption. Go right ahead. I'll just keep taking notes here." One of which was to bring my micro cassette recorder to our next meeting—if we had a next meeting—my fingers were starting to cramp.

"Good idea. A tape recorder would be helpful," she said.

"Will you quit that!"

"Sorry, habit. Anyway, another view supports the theory that life on Earth may have originated on another planet or star, the most recent example being Mars, and was transported to Earth by a meteor. But, just like the Superstring Theory that doesn't explain where that life came from in the first place. It is possible that life started on another piece of space matter and was transported here. So, did life start on Earth some 3.75 billion years ago? Or, did life start elsewhere and come to earth? Perhaps there is more than one form of 'life' on Earth. The form we are aware of and another one or two or ten or twenty forms still buried deep in the Earth's core or under some thick icecap. Did what we know as life on Earth today come from the original life form or was the original wiped out by the first great extinction 425 million years ago—or one of the other four major extinction's that have occurred since then—and the present form of life came after? As you can see, there is a myriad of possibilities as to the origin of life on Earth."

"So which one is the correct one?" I asked.

"I don't know."

"You, Mother Nature, don't know?"

"Nope."

"Imagine that," I said.

"Here we go again with the sarcasm. I guess you think that if I were *really* Mother Nature I would know that?"

"Bingo." *May as well throw in some smugness with the sarcasm. I'll take my fun where I can get it.*

"Why?"

"Why? Because you're Mother Nature."

"Right, Mother Nature. Not Mother Origin of Life and, certainly, not God."

"Oh, I see. So only God knows the origin of life."

"Exactly. Oh, and that's where the expression came from, by the way," she said.

"Huh?"

"The expression. 'God only knows'."

"That's rich, Mona" I said. "Now let me get this straight. You and your crew arrive on a planet and spend billions of years getting it ready to support life, but you don't know where that life comes from or how it begins. Do I have the gist of it?"

"You do indeed."

"I don't get it. How do you know what to do?" I was truly flabbergasted.

"I just do. I can't really explain it. Haven't you ever known someone who could build things from scratch without any plans or blueprints? They just have that innate ability. They just know how to do things?"

"Sure," I answered. "My brother David is like that. He is great working with his hands doing carving, woodworking, and mechanical stuff."

"It's the same with me when it comes to my work. I don't remember learning it. I don't know where the ability came from—although I have a good idea. It's just always been there.

Others on my crew relate the same experience. We go to a solar system, do our thing, and move on to the next assignment."

"You stated earlier that you had made some mistakes. Green skies, pigs with wings…the platypus." She shot me a warning look and I held up my hands in mock surrender. "How do mistakes happen with God involved? I've always heard that there are no mistakes in God's world."

"Semantics," she answered. "My understanding is that God—or, again, whatever—sets the whole shebang in motion with the component necessary for life already in place. When the time is right, my crew and I are dispatched to that region to do our collective thing. Sometimes, not being God, one of us gets our calculations a little wrong or plants a seed where we shouldn't have and viola, pine trees in the desert."

"Then, again, mistakes," I persisted.

"Then, again, semantics," she persisted. "You circle the wrong answer on a test when in a hurry even though you knew the correct answer, you call that a mistake. I plant a seed in the wrong location, I call that a mistake. But, in reality, who are we to say that our actions were truly mistakes? Maybe subconsciously the actions we took helped us get to where we were supposed to go or helped something or someone else get to where they were supposed to go. And, without our mistakes that wouldn't have been possible. Maybe mistakes are just the soul's way to guide us, get us back on track. None of us knows to a certainty what is going to happen from one moment to the next. We may be able to predict an event to the millionth decimal point, but there is always that one in a million chance that the event won't come off as predicted. In fact, your quantum mechanics, through indeterminacy, has shown you the fallacy of thinking you will ever be able to explain exactly how and why things work the way they do."

"What is indeterminacy?" I asked.

"Indeterminacy means that even though the law of averages, etc. can make predictions like those I spoke of earlier—that things on a large scale are very predictable—on the quantum level there is a randomness, an unpredictability that makes exact behavior, and therefore the exact nature of things, unpredictable. The universe and everything in it is constantly changing and we can't predict with certainty what those changes will be or the resulting outcomes of the changes when they do occur. So, through evolution, natural selection, and the like, Earth and its inhabitants have proceeded to their present state. And, what a state it is in I might add."

"And, God just let it happen," I said. "That's hard to believe."

"You are hard to believe, if you believe that, believe me," she said.

"Huh?"

"I said," she started to explain, but I cut her off.

"Never mind. I get the idea."

"Oh, okay. As I was saying, free will, the power to choose is among your greatest gifts. The most destructive, but one of the greatest nonetheless. So, in that respect, yes, God 'just let it happen'."

"So, you truly don't know how life originated?" I asked. She looked at me like I was slow on the uptake.

"No."

"That's a convenient out."

"An out or not. It's the truth. I figure if God wants me to know that then God will tell me. Until then, I will keep doing my job and not question how God does his."

"I must admit, you cover yourself well."

She ignored that and said, "Maybe we should switch our focus and look at how we can preserve life instead of worrying about where it came from in the first place."

"But, don't you think that if we knew the origin of life we would have a better chance of preserving it?" I asked. "After all, you are the one who put the emphasis on history."

"Perhaps knowing the origin of life would help in its preservation. Maybe it would help in finding the cure for cancer or AIDS or Alzheimer's Disease or all of the other diseases rampaging their way across your planet. But, even if you were to discover the origin of life today, I don't believe that having that knowledge would help you purify the water or clean the contaminates from the air or close the hole in the ozone layer. You would still need to deal with the environmental issues that threaten to make this planet uninhabitable for human life—and a lot of other life forms, too—within the next 50 or so years."

"Then again," I said. "Maybe we could use the knowledge to begin life anew."

"It has taken 3.75 billion years for that origin of life to evolve to its present state. Even if you discover the origin of life tomorrow it may be a little late to use the information to help with today's problems. So don't you think it might be a good idea to try to halt the destruction of the planet just in case?"

"I guess, but I still think it would help to know it."

"Then have some of your scientists continue to search for it, but urge the rest to shift their focus to environmental concerns."

"Easier said than done, I'm afraid," I said.

"Then motivate them," she replied.

"How? With what?"

"Money, modern man's greatest motivator. Or so it would seem."

"Ah, yes. Money," I said.

"The universal language in cultures such as yours."

"Gee, and I thought love was the universal language."

"In enlightened cultures. Unfortunately, yours is not an enlightened culture with the exception of a very few people. A survey of your species has them saying that religion is the primary influence on their behavior. Personally, I believe its money."

"Ouch," I said. "That's rather harsh, no?"

"Harsh, but, true. Fortunately, you can turn that into an advantage. There are already quite a few technologies that are ecologically friendly and highly profitable at the same time. In your culture, that should be motivation enough, don't you think?"

"Absolutely. 'To make nice green make nice to green'."

"There you go. You already have a slogan."

"Cool. I can see it already. T-shirts and ball caps, bumper stickers and coffee mugs. Hey, a Web site. This could get very lucrative." I was jazzed.

"As I said, a great motivator." She could do sarcasm, too.

"Exactly, as you said."

"Just so you don't distort the meaning behind the slogan," she retorted.

"No problem," I said. "And, speaking of 'meaning', that leads me to my next question."

"And that is?"

"Since God hasn't told you the secret of the origin of life, I'm guessing that he hasn't told you the meaning of life either?"

"Of course not."

"No, of course not." *Another convenient out?*

"Why would she?" I smiled at her change of pronoun. She returned my smile before continuing, "Perhaps that was why she gave us free will, freedom of choice. It leaves it up to each individual to decide. In fact, I think a better way to phrase the question would be 'what gives your life meaning?'. When do you feel most alive, most fulfilled, most useful, most productive, most a 'part of'? Maybe life is just a process in which to experience and learn things and pass them on to others. Some humans—and some other animals such as dolphins, elephants, and primates—seem to get their greatest pleasure from helping others, passing on their knowledge and wisdom. Maybe it's in the pure joy of discovering new things, having new insights, those moments of 'aha, I get it!', the thrill of acquiring knowledge. For you it may be one thing, for another, something else entirely. Maybe it's a way for God to experience the 'earthly'? Or, to experience itself in different ways. Maybe the purpose is to just keep evolving. Bottom line, maybe each individual has to figure it out for themselves."

"Maybe you have something there," I said. *Crazy or not, her diatribe made some sense.*

"Thank you," she said. "I know it's not easy for you to admit that a crazy lady might just have a valid point."

"This is true," I said while flashing her a sheepish smile.

"So, what gives your life meaning, Michael?"

"Good question."

"Thank you. Do you have an answer?"

"I'm not sure. Give me a moment"

I put my pen down and closed my eyes. I thought about what she asked. As for relationships, Mona was correct. I've never been married nor had children. My life has consisted of a series of short-term encounters interspersed with long periods of time alone. What I had with Lily at present was no different. We have been apart more than we have been together. I have infrequent contact with my family, usually restricted to birthdays and holidays or the occasional e-mail. I e-mail or talk on the phone to friends, also, sporadically. We occasionally get together for dinner or at a party of some sort. As I said before, I pay my bills by running a computer system in a factory—not exactly my dream job. If you have ever worked in a factory or seen the inside of one, you know what I mean. They are dark, oppressive, noisy, and negative. To be brutally honest, it's a soul-killer. I usually spend my free time at home reading or writing or watching the occasional TV show or cable movie or playing with Nikita, my cat. For exercise I run and bike and sometimes lift weights. What *does* give my life meaning? The knowledge I gain from reading? My writing? Working out? Playing with a cat? I had to admit I lead a fairly solitary existence. Maybe writing this book is just what I need to give my life some meaning, a purpose. I opened my eyes and looked at her. She had a knowing smile on her face. Which kind of ticked me off.

"I guess you 'heard' all of that?" I asked.

"Yep."

"So, I guess you also know that I'm hooked into this now regardless of whether or not I believe you are Mother Nature?"

"Right, again." She could do smug, too.

"I figured as much. And, you knew that would be the result when you asked me what gives my life meaning, didn't you?"

"You're on a roll." Her smugness had given way to a friendly smile.

"Pretty sneaky, Mona." I had to smile myself.

"I have my moments."

"That you do." *Conned by a crazy woman. And, I had just agreed to work on writing a book with her. Maybe I'm the one who's nuts.*

She laughed. "I believe that I can calm your anxieties." She pulled the medallion that I noticed her playing with before out from under the top of her dress. She held it out for me to look at. At first I didn't know what I was seeing. The image in the medallion appeared to be moving. I was drawn into it; I couldn't pull my eyes away. Immediately a kaleidoscope of nature scenes flooded my vision and I had the sensation of leaving my body.

First, I experienced a dense tropical rainforest lush and green. It was alive with monkeys frolicking in the trees, gorillas and baboons playing on the forest floor. Cheetah and leopards stalked lemurs and chimpanzees. Beautiful flamingoes waded in tropical ponds. Flocks of yellow orioles and groups of colorful butterflies filled the air along with swarms of tsetse flies. Macaws perched on tree limbs as tree frogs made music from below. Beautiful floral species such as the passionflower, the rambutan, and the hericonia grew along side the cashew and banana trees. Avocado, coffee, and cocoa plants were in abundance. Cobra snakes and vipers slithered through the bush. Rhino and elephant could be seen at the forest's edge. Great lions hunted a herd of zebra while the hyena waited to scavenge the leftovers. An indigenous tribe of Kogi hovered around the carcass of a water buffalo that they had slain during a hunt, blessing it for its sacrifice in providing them with food and shelter.

From there I was whisked away to the coastland and witnessed sandpipers doing their schizophrenic dance at the waters edge. The white sands of the beach were littered with a plethora of seashells in all sizes, shapes, and colors. Flocks of sea gulls and pelican glided above the waves to scout for food. I submerged beneath the water and swam with porpoise and bottle-nosed dolphin and humpback whales. They kept me safe from the predatory mako and hammerhead sharks and the sleek barracuda. As I went deeper, I passed squid and blue-ringed octopus and manta rays. Beautiful coral reefs were overrun with schools of kelp rockfish, sunfish, trigger and clown fish in a rainbow of colors. Scattered about the ocean floor were sea cucumbers and anemone, oysters and clams, sea urchins and plankton.

The sandy seabed was turned into an inferno as I found myself in the desert of the American Southwest. The sun beat down from a clear blue sky on prairie dogs as they darted in and out of their holes. Brewer sparrows alit in Joshua trees, creosote bushes, and desert ironwood. King snakes slithered around the base of giant Saguaro cacti that cast shadows resembling some prehistoric pitchfork. Yucca and ocotillo plants grew in abundance next to beautiful dune primrose. Horned lizards dodged sagebrush as it skittered across the sun baked desert floor.

The desert gave way to the Great Plains where fields of wheat and grain and barley gently swayed in the summer breeze. Sparrows flew above the fields keeping an eye out for the great hawks that stalked them from above. Pheasant and sage grouse nested in the bunch grass. White tailed deer and great bison grazed nearby.

Next, I was swept away to a majestic mountain range of snow capped peaks where mountain goats skirted their way along the precarious cliffs. A bald eagle gracefully circled above before swooping down to settle on the branch of a huge Ponderosa pine. Stands of spruce and Douglas fir shared space with juniper and lady fern and mountain heather. Sleek lynx and leopards stalked the wilds in search of dinner. Mule deer and elk populated the grasslands at the mountain's base. Red-tailed hawks glided through the treetops on the lookout for a badger or wolverine in the bracken below.

I floated above a crystal clear mountain lake that was brimming with catfish and sturgeon, gar and bass, and huge carp that brushed aside the cattail as they lumbered their way through the water. Water lilies floated on the surface bumping into the willows at shore's edge. Frogs and toads made throaty music when they weren't snagging flies and mosquitoes from the air. Mallard swam on the lake as a flock of geese flew by overhead in their familiar 'V' formation.

I became a part of a river that flowed from the mountain, swollen with salmon and bass, into a forest overrun with stands of Sitka spruce, red cedar, aspen, and sugar pine. Clumps of shrubby penstremun and Pacific rhododendrun were home to scrub jay and western bluebirds. Hummingbirds and woodpeckers filled the forest with their strange sounds. Beautiful bouquets of Washington lilies dotted the landscape. Lovely white pine butterflies flitted through the air coming to rest on cliff rose. Grizzly bear hunted salmon at river's edge. White bobcat and Grey fox stalked red squirrel, raccoon, and bush rabbit.

Regretfully, I surfaced from my trance amazed by what I had seen, but more amazed by what I instinctively knew. I had just been given a glimpse of what the earth had looked like thousands of years ago when it was still inhabited by people who lived in harmony with nature. And, I knew that Mona was who she said she was and that it was vitally important that I listen to what she had to share with me and pass it on the best I knew how. My amazement was soon replaced by a conviction that I was exactly where I was supposed to be at that point in time and that my life did in fact have meaning. I was overwhelmed with gratitude. I looked at her and said, "That's some necklace."

She laughed and nodded in agreement.

"So, why didn't you just share that with me from the get go? It would have saved a lot of hassle, no?" I asked.

"I have no doubt that it would have, but I thought it best that you make the decision to be a part of this project of your own volition. That way we would both be sure of your, dare I say it, commitment."

I had to laugh out loud at that one, as did she.

"So, now that I am committed," I said as I faked a shudder which drew another peal of laughter from her, "how about showing me how to do that 'intuiting' thing whenever I want to?"

"As I told you before, you will get there on your own through the actions you take."

"Like agreeing to this project?"

"Like agreeing to this project," she concurred.

"But, you could speed up the process so I could be there say, now, by doing something like you did with the nature visions, right?"

"Yes, I could."

"Then why won't you?" I asked.

She shook her head slowly, distractedly and said, "My husband wants to address that one." Her voice changed to a deep, rich, booming baritone. "Typical of the human race, the modern version at any rate, to grasp for the quick fix, to yen for instant gratification. That's one of the reasons this planet is in the shape its in." Then, as quickly as she went away she was back.

"That was creepy," I said, shuddering for real this time. "You could have just translated that, you know."

"True, but that wouldn't have been near as dramatic...or as effective."

"Or, as scary. Jeez, I thought you were channeling Darth Vader"

"Got your attention didn't it?" she asked, merriment filling her voice.

"I'll say. So, that was Father Time?"

"Yes, that was my hubby." The way she said that convinced me she was very much in love.

"He seems a bit, uh, harsh." I didn't wish to offend her. Or him.

"I told you that he's not much of a people person. But, he is really quite harmless. Trust me."

"I do," I said. And, oddly, I did—very much so.

"Good. Shall we continue?" she asked.

"Why not," I answered.

"Yes, why not. I suppose I should digress a bit here to the beginning of your solar system."

"Groovy." She gave me a blank stare. "A reference to the 60's. Get it?" I asked.

She nodded patiently and smiled, then continued, "We already know about the Big Bang from our earlier discussion. Again, that happened approximately 16 billion years ago. To cut to the chase, your Milky Way galaxy was formed approximately 7.75 billion years ago, which was about the time my crew and I arrived. By the way, why did you name your galaxy after a candy bar?" It was my turn to give her a blank stare—and it felt childishly good to do so.

"Sorry," she said. "Just trying to follow your example and lighten it up a bit. Anyway, about four billion years later, right after my crew and I left, is when the Big Bang theory postulates that life in the form of bacteria and algae appeared. We already discussed this in its entirety so I won't rehash that now unless you have any questions."

She paused to let me answer. I shook my head 'no' and flexed my fingers.

"Good. Evolution continued at its excruciatingly slow pace— by human standards, at any rate—producing multi-cellular animals, aquatic life, winged insects, small land animals. Then, 425 million years ago, the first major extinction occurred during what your scientists call the Ordovician Period, one of your Ice Ages. That extinction wiped out pretty much everything but some of the skeletal creatures that were around at that time.

"The survivors of that extinction continued on their evolutionary way and weathered the storm of another extinction at the end of the Devonian period 365 million years ago; another major event occurred around 225 million years ago, theorized to have been an event of massive volcanic activity, which killed off 90% of the living species at that time. This is the extinction event to which I referred earlier. At the end of the Triassic Period, around 190 million years ago, a meteor struck at what is now your Quebec's Lake Manicouagian which resulted in yet another evolutionary weeding out process. Finally, 65 million years ago the last major extinction took place—excluding the extinctions happening continually on a daily basis, which I happen to consider major—and, of course, the extinctions to come of your current ecosystems if something isn't done. This most recent extinction was the result of dual impacts by meteors in the Chesapeake Bay area in North America and in Siberia. This is the extinction that you have heard so much about that wiped out the dinosaurs and allowed the smaller mammals to evolve into what were the first humans around three million years ago."

"Dinosaurs are cool."

"Yes they are…or were. But, no more or less so than any other species past or present."

"Oh, come on. You can't tell me you think a creepy, filthy cockroach is as cool as a big old dinosaur." I was serious, but I also thought this might get her goat. As I said, I take my fun where I can get it.

"Yes, I can," she said in a very deliberate tone of voice. I kept my face slanted down towards my legal pad so she wouldn't see me stifling my laughter. She continued, "I don't feel that one species is any cooler than another is. As you will learn, all species are cool in their own way, have their own unique function to perform, and serve—or served—a useful purpose. Everything is a part of the 'all', part of the balance."

"Okay, I see your point. But, I still think dinosaurs are cooler than cockroaches."

"Only in body temperature," she shot back.

"Because the dinosaurs are reptiles?" I asked.

"No, because the dinosaurs are dead."

"That's cold."

"You're right," she said. "All this talk of extinction has put me in a foul mood. I think I need a change of scenery. Let's go outside."

"Sounds like a plan," I said.

I gathered up my belongings and we adjourned to the outdoors.

CHAPTER 3

BACK TO THE PRESENT

It was a beautiful spring day: temperature in the low 70's, a mostly sunny sky, and a light breeze. I followed as Mona led the way through the parking lot to the park adjacent to the library. Like the library, the park was nearly empty with the exception of a few of the area's homeless population who were huddled in a small group by one of the many available park benches. Mona eschewed the benches, preferring instead to sit on the ground with her back to one of the few oak trees that hadn't been cut down yet. I sat facing her and retrieved my pen and pad from my pack.

Mona took a deep breath and coughed. "Excuse me. I'm not used to breathing air this foul. It's an irritant—in more ways than one. Still, it is wonderful to be outside on a day like today, or on any day really."

"Yes, it is," I agreed. "I love being outside. My dream is to live in a warm weather climate, Arizona, New Mexico, somewhere where I can be outside year 'round without having to dress like Nanook of the North."

"Well, if global warming continues on its present course unabated, you may not have to move at all," she said. "It will be plenty warm here in Ohio year 'round. But, I digress. We will get to global warming soon enough, but for now we need to stay focused on the history lesson."

"Why are you spending so much time on history?" I asked. "I mean, the problem is what's happening now, right?"

"Right, but I feel that if you don't know how things got to be the way they are, it will be much harder to understand how pressing today's problems are and why time is of the essence. And, most importantly, to see that there are solutions to the problems. You'll just have to take my word for it."

"Okay, it's your show. I believe that you were telling me about the extinction event that killed off those cool dinosaurs and paved the way for mammals, especially of the humanoid variety, to evolve."

"Right. And, just look at how you humans have 'evolved'." She pointed to a piece of litter blowing through the park. It was the wrapper from a fast-food hamburger. "How appropriate. That one small scrap of paper represents so many of the larger issues that your planet is facing: deforestation, air pollution, solid waste dumps, chemical contamination of water sources, land mismanagement, diseases, and specie extinction." Her eyes took on a look of profound sadness.

"There are a lot of careless people in the world," I said.

"Yes, there are. Careless and ignorant. And, I don't mean from lack of brain cells. I mean ignorant as in not aware of the problems. Then, again, some *are* just thoughtless. They just don't think about the consequences of their actions or if they do, they make excuses and do what they want to anyway. Your society is so self-absorbed you can't see the forest for the trees—of course, that could also be because you have cut most of them down in the name of 'progress'. Vast majorities of you humans seem to have gotten it in your heads that evolution stopped with you.

The arrogance in that belief is incalculable. How could you possibly think that every other life form on the planet is still evolving—bacteria, viruses mutating and evolving so fast medicine can't keep up with them, insects that are developing resistance to the insecticides you try to kill them with, to name a few—but humans have ceased to? You think that you evolved from the same tree as the ape to Homo habilius to Homo erectus to Homo sapiens then just stopped evolving? You don't think that Homo sapiens can be improved upon? Look around at the destruction Homo sapiens have done to the planet, to the other life forms on the planet, and to each other and tell me that you are at the apex of development, that you are so evolved that you can just maintain the status quo."

I started to interrupt, but she held up her hand to stop me. I figured this might be a good time to keep my mouth shut.

On she raged, "Why is it so hard for your kind to get your collective mind around the idea that just maybe you are still evolving and that just maybe you are to show the way to the future generations of your species and on down the scale to the so called lower species. Maybe you are just going to be the first species to reach the next rung on the evolutionary ladder while the chimpanzee or the dolphin moves up to take your place. Maybe the meaning of life is to lead the way by example. And, if so, what is your example going to be? Destruction of life as you know it on the planet or its restoration and continued evolution? Because mark my words, you have to choose very soon what your legacy is going to be—you are making the planet uninhabitable for life as you know it. You aren't killing the planet as some people scream—you can't 'kill' the planet, although it seems that you are certainly trying. The fact is that the planet will one day simply shrug you off and regenerate itself. But, that means destroying most if not all life as you know it so it can accomplish that. Is that what you want for your children and grandchildren?" We looked at each other for a moment. I assumed she was asking a rhetorical question since I didn't have any children. "Well?" she asked. My mistake.

"No, of course not," I answered with conviction. I hoped it was convincing anyway. She seemed to be getting angry, again.

"Forgive me," she sighed. "I was hoping that coming outdoors would brighten my mood, but it seems to have just worsened it."

"No need to apologize," I said.

"It's just such a shame. Look at the beauty all around you." She pointed to a cloud formation that was floating past us, to another of the few trees left in the park, to a pair of squirrels that were chasing each other through the grass. "You are going to destroy all of this because you just don't appreciate it. You take it all for granted and that is going to be your downfall. It's all so unnecessary, so pointless. It makes me so sad."

"Some cheese with that whine?"

"Pardon me?"

"You're a whiner," I said. "In fact, I'm going to start calling you Mona the Whiner. 'You're going to destroy all of this. It makes me so sad.' Waa, waa, waa."

Her eyes flashed black, but only for a moment (during which time I thought that maybe I had made a big mistake), and then they began to sparkle. She threw her head back and started laughing. The sun actually seemed brighter, the sky bluer, the air cleaner. She caught her breath. "And, you are a real pip."

"Gee, thanks, Mona."

"No thank you, Michael."

"For what?"

"For the humility check."

"Anytime." *Yeah, like that had been my intention.*

"Sometimes I take myself too seriously," she said. "It puts the spotlight on the messenger at the expense of the message. Big mistake. This is too important."

I gave her a holier-than-thou look and said, "I'm glad you realized that. There is no room for egos here."

"Oh, jeez. Don't push it, Michael," she said.

"Hey, I'm just another of your wonderful creations," I retorted.

"So much for no egos."

"Touché."

"You ready to get back to your history lesson, oh wonderful creation of mine?"

"Sure."

"Good. As I was saying, the extinction of your cool dinosaurs did indeed open the door for the small mammals that had survived the extinction event to evolve. Everything was proceeding along nicely. The extinction events and what your Darwin called natural selection were keeping things flowing as intended, naturally. All of the extinction events that I have mentioned up to now were the result of naturally occurring events—the Ice Age, meteor impacts, cosmic disruptions, volcanic activity. What is happening today is the result of unnatural events. By the way, I use the terms 'natural' and 'unnatural' in the same way I do God—because it is the easiest way for you to understand what I am getting at. Humans are a part of nature and therefore by definition what they do is natural, but in this context we are talking about events that would not have taken place were it not for the actions of humans and humans alone. Because of deforestation, pollution, the population explosion, and the like—all human caused events—species are dying out at an alarming rate. The majority of these modern day extinctions are occurring in the rain forests around the globe, especially on the continent of

South America. The reasons are complex and the consequences are many; some already showing themselves, some of which have yet to be felt. I will get into this in greater detail in a bit."

She plucked a blade of grass and began to absentmindedly tie it into knots. I noticed that the tree she was leaning against seemed to be cradling her as if it had conformed its shape to accommodate hers. I thought at the time that it was just my imagination, but as I write this, I am not so sure. In light of the morphing of her appearance that I had witnessed earlier and other things that happened over the course of our dialogues, it would not surprise me in the least if it really happened as I now remember it.

"As the evolution of the hominids progressed through the stages mentioned previously," she continued. "These early beings developed a lifestyle that was tribal in nature. They lived a day at a time in what today are called climax communities. These are communities that use up only as much energy as they create. They didn't hoard food or strip an area totally of its resources. They stuck together and cooperated in hunting, food production, child rearing, health care, and maintaining the integrity of the community. By living in harmony with their environment and realizing that they were just one spoke in the wheel of nature, they lived relatively peaceful, stress free lives. They took what the environment provided, in moderation, and let other tribes do the same. They were the first environmentalists."

"You make it sound so wonderful," I said. "But didn't they have it really hard? I mean, no electricity, no running water or plumbing, no telephones."

"They had no concept of those things so they didn't miss them. From their frame of reference, life wasn't judged to be easy or hard—it was just life. Overall, they were happy and content with their lives."

"But, still," I said. "No bathrooms or refrigerators or microwaves."

"Yes, imagine that. But, you left out a few other things they didn't have either."

"Yeah, I know," I interrupted. "Cable TV, computers, automobiles, air travel." I paused to think of some other examples. She had a few of her own.

"Cancer, world wars, genocide," she said.

"Oh, come on."

"I'm serious. Cancer is basically a man-made disease. Only 10% of the cancers are caused by genetics. The other 90% are a result of environmental abuses. As for world wars, the majority of the tribes followed the Law of Limited Competition that states that you may compete to the full extent of your capabilities, but you may not hunt down your competitors or destroy their food or deny them access to food. You compete you don't wage war. Just like in the animal world, if there was a dispute over territory or resources, the resident attacked, the invader withdrew. Occasionally, but only enough to remind each other that individual communities were not to be interfered with, one tribe would attack a neighbor. But, this happened rarely and was only done with enough force to get the point across. The purpose was not to conquer, enslave, or annihilate."

"Sort of a 'don't get any stupid ideas about poaching or we'll kick your ass' reminder," I said.

"To put it rather crudely, yes. This lifestyle creates and maintains a beautiful, natural balance: the interaction between 'food' and 'feeder' populations. It keeps the population of all species in check naturally. When food sources dwindle, the species in the area, including humans, naturally adjust their output of offspring to coincide. The population of the species in areas such as these varies from 95% to 105% of optimum capacity. It grows and shrinks according to food availability and the birth/death cycle. It doesn't continue to grow unchecked, like it has over the past few hundred years, to the point of over-saturation. This system is known as a negative feed back system: as the population outgrows the sustain-ability of an area—in other words, consumes food to the point where food becomes scarce—the population declines until stabi-lization occurs. By the way, your current rate is 2.5 million births versus 1 million deaths per week, which is way out of balance.

"On most occasions, symbiosis would occur between the peo-ples of different tribes, again like in the animal world. Symbiosis is the joining together of separate organisms to deal with adverse situations for mutual benefit. You see symbiosis all the time in nature: in the seas, with cleaner fish eating the parasites from the mouths of sharks and on the African plains, where gazelles eat the grass, giraffes eat from the tree tops, lions eat the weak of the herds to keep populations in check."

"Fish that eat parasites from the mouths of sharks. Yuck," I said.

"Yeah, yuck. Much grosser, I'm sure, than eating monkeys brains or beef tongues or veal."

"Yeah, okay, point taken thank you very much," I said while trying not to gag. "Please continue."

"Tribes lived for almost 2 million years this way without harming the environment at all. They trusted God, the process, Mother Earth, whatever, and the lifestyle worked. There are some examples of this tribal way of life that have managed to survive to this day: the Kogi in South America, the African Maoris, the Penan of Borneo."

"Doesn't the fact that this lifestyle has been pretty much elim-inated mean that it wasn't an effective way to live?" I asked.

"Hardly. Just because the tribal way of life has been pretty much eradicated from the planet doesn't mean that it didn't work. It still works in the areas where people are left in peace to follow that way of life, as I just stated in my examples. The fact that it isn't practiced today any more than it is doesn't mean it was a casualty in the process of 'survival of the fittest' or natural selection."

"Would you explain that, please?" I asked.

"Certainly. A prime example is the Aborigines in Australia. Before the Europeans arrived a few hundred years ago, the Aborigines lived a relatively peaceful life in a structured, spiritually based society that left them plenty of leisure time. They existed on a nutritious, diverse diet consisting mainly of vegetables and fruits and supplemented with the occasional meat kill. They had developed no technology, not because they were ignorant savages, but because they had no need or desire for it. They were happy in their lifestyle. The Europeans, in contrast, were technologically advanced, especially in the area of weapons, and brought firearms to the continent. With this superiority of war making, the Europeans subdued the 'heathens' and tried to educate and socially and religiously reform them. When this didn't work, they resorted to such tactics as hunting them for sport and poisoning their food supply. Finally, the Aborigines were ostracized and forced from their ancestral lands. Presently a vast majority of them live in squalor and are plagued with crime, poverty, disease, unemployment, and alcoholism. Through it all, they have tried to honor and maintain their tribal ways. Sound familiar?"

"I'm ashamed to admit it, but yes," I said. "Their plight is similar to the way the Native Americans were treated when our European ancestors arrived in the Americas."

"Exactly. And, as you can see, what happened to the tribal way of life was far from natural. Genocide is not natural. It defies everything that God and nature created and stands for—hence, the chaos on your planet today. And, speaking of Native Americans, the last full-blooded Kaw Indian died this past spring. Meaning that tribe of people is now extinct."

"Why do we humans do things like that if it is, in fact, against our nature?" I asked.

"Because you shifted away from a spiritual based life to one based in materialism. If one event could be pinpointed as the catalyst for what ails your planet today it would have to be the agricultural revolution of about 10,000 years ago that first occurred in the Middle East in the area known as the Fertile Crescent. The people in that area figured out that if they planted crops in abundance and stored the excess food for longer periods of time they wouldn't have to move around so often. This was the beginning of the first organized, permanent communities. This signaled the start of the shift from the tribal lifestyle—the spiritual based example that I mentioned a moment ago—to what I call the hoarder culture, a more materially based way of life. As the hoarder lifestyle gained in popularity and spread to other areas, the Law of Limited Competition and the attitude of 'Live and Let Live' was also abandoned and gave way to a fear based gestalt. This fear was manifested in many ways. One of which was the thinking that anything or anybody different was bad or wrong and must be brought around to your way of thinking and doing or be destroyed. This justified your expansion and the genocidal behaviors that I gave you examples of with the Aborigines and the Native Americans.

"The next link in the chain of events was that the hoarder culture changed the paradigm to one of positive feedback: when the population of an area consumed food to the point of scarcity, they would conquer more land to plant more food and the population would continue to rise. In other words: more food equals more people; more people equal the need for more food. This cycle of positive feedback has continued unabated for centuries now, creating a huge population explosion resulting in mass starvation."

At this point, she looked across the park at the homeless people. I followed her gaze and for the first time really looked at those people and saw...people. I felt a sense of shame at how I had always ignored their plight. I turned back to find Mona watching me and knew that she had intuited my discomfort.

"So what do we do?" I asked.

"The answer to overpopulation isn't to produce more food," she answered. "The answer is to stabilize food production and give the population a chance to stabilize."

"Wouldn't that just mean more people starving?" I asked.

"No. You have plenty of food; it's just not getting to where it's needed."

"That doesn't make sense to me," I said. "If there is plenty of food and it's not getting to the people who are starving, then where is it going?"

"It's going to feed the expanding population of the 'have' cultures, the cultures that can afford the food. The rest is hoarded, kept under lock and key. Fully 80% of all malnourished children live in cultures with food surpluses. The families just can't afford to buy it."

"But, even if we got the food to the starving people who can't afford it, wouldn't that cause an even bigger increase in the population?" I asked.

"Not if you only distributed enough to feed the hungry with nothing left over for hoarding and quit producing excess food in the 'have' cultures. In fact, if you would only produce enough food for 95% of your world population for a decade or two, you could not only stabilize the population, but also cut down on the obesity problem in the 'have' cultures at the same time. In your country, 23% of the population is considered obese, including 20% of the children. So you have one in five kids eating to the point of obesity while at the same time another one of those five kids is suffering from malnutrition. Does that make any sense to you?"

"No, it doesn't. In fact, it's obscene," I replied, horrified.

"A policy like the one we just discussed would also have a positive effect on the environment, which I will elaborate on shortly."

"That would work?" I asked.

"It always has in the past," she responded. "Study after study has been done with populations ranging from mice to humans that prove this out."

"Maybe we should do that then."

"You think?" She shook her head and went on, "Anyway, over the next 4,000 years agricultural technology continued to advance. Communities grew as the first population explosions occurred. Again, if you store food and create an over abundance, you create perfect conditions for a population explosion. Until the lifestyle change from tribal to communal, populations doubled every 19,000 years. In the past 10,000 years alone, the population has multiplied 60 times to a total of 6 billion people. This cannot continue without severe and some believe, fatal, consequences."

"Fatal?"

"Fatal."

"That's not a good thing," I said.

"Depends on your perspective. I'm sure humans would not view it as a good thing. But, as for the other life forms on the planet, my guess is that they would view it a bit differently."

"Your guess is probably pretty accurate," I mumbled, which earned me a chuckle from Mona.

"To proceed," she said. "The hoarder lifestyle continued to strengthen its hold with the advent of the ruling class, the rich vs. the masses, the poor. You humans began to believe that you were the only creatures with souls, that only you were 'created in the image and likeness of God' instead of realizing, since *everything* came from God, *everything* is 'created in the image and likeness of God'. This secularization from nature allowed you to rationalize its subversion to fit your needs. The natural world used to be viewed as an extension of the self, so the destruction of nature was seen as an act of self-destruction. Cultures such as yours got away from viewing the natural world in that way. As a result, your morals were compromised to the extent that you claimed the right to choose what lived and what died and even *whom* lived and died. Superstitions and sacrifices blossomed from another of your fears, the fear of the unknown. What had once been a stable way of life, the tribal life, the known, became unstable urban life, the unknown. Ruling classes continued to gain power; war and slavery became the norm. The idea of suffering and the need for salvation flourished in this atmosphere as the tribal lifestyle fell further out of favor and the tribes themselves fell victim to genocide and slavery. This climate also gave rise to the worlds major religions: Judaism, Hindu, Christianity, Buddhism, Islam, and Muslim."

"It seems that the religions would have led us back to a more spiritual way of life," I said.

"You would think that, and in fairness, maybe that was the intention of most of them when they began. After all, the message of the various Prophets that the majority of these religions are based on was one of peace and love and tolerance, brotherhood and compassion and non-violence. But, the message of these Prophets was twisted and bastardized to one of separation and intolerance and close mindedness. Anyone who doesn't believe in a particular set of tenets set forth by their particular church is bad and sinful and needs to be punished. If you don't do things their way, which they consider to be the only right way, you go to hell. God forbid, pun fully intended, anyone is different.

"Out of this attitude came the Crusades in the 12th and 13th centuries; the Inquisitions of the 14th and 15th centuries; the religious persecution in Europe that sent many of your forefathers to North America in the first place, who ironically carried out their own persecutions with the witch burnings in Salem, Massachusetts; the Catholic/ Protestant strife in Ireland; the ethnic cleansing in Bosnia; the Buddhist exile from China to Tibet; the Jews in Europe, especially in the 20th century; the Muslims in the Middle East; the Jews and Palestinians. A great many of you humans are very closed minded. Its no wonder that the mind set and group consciousness that arose from these abominations produced the likes of Attila the Hun, Ivan the Terrible, Mussolini, Hitler, Stalin, Pol Pot, Hussein, and the like."

"We are a strange species," I said.

"Another of those colossal understatements, I would say. But, we have gotten a little ahead of ourselves. The first such 'civilization' to prosper under the agricultural lifestyle was the Sumerians in Mesopotamia, in the Fertile Crescent that we spoke of before. They ruled for 4,000 years until they were destroyed by their abuse of nature. They destroyed their lands and all the lands they conquered through deforestation and reckless agricultural practices to feed their ever-expanding population, which eventually led to poverty, famine, and mass starvation. Today, that once lush area of the Earth is mostly a desert.

"The Greeks followed and built one of the most incredible societies ever known to man, but eventually succumbed to the same fate as the Sumerians. They ruled for 1,400 years until their abuse of nature did them in as well.

"The Romans, learning nothing from their predecessors, ruled for 1,100 years until they too were destroyed by the same practices."

"In fairness," I said. "There were no newspapers, television broadcasts, or Internet to analyze and communicate the events of history at that time and get the relevant information to the people."

"Like that would have made a difference," Mona replied. "You have those outlets today yet very little gets changed."

"Oh, come on. Everyday I read something in the papers or see a newscast about the environment," I said. "And, I know that there are tons of web sites on the Internet that address these issues."

"Yes, you get an obligatory article or broadcast occasionally pointing out yet another ecological disaster and that's the problem. Most of these news items appear *after* the fact. Anything predictive about ecological problems is treated with disdain as if the predictors were so many Chicken Littles saying the sky is falling."

"Or has a hole in it."

"Or has a hole in it, yes. For every report or web site dealing with environmental issues in a sane, objective, truthful manner there are a dozen nay-sayers crying 'conspiracy' or calling them 'alarmist'. And, most of those naysayers are sponsored and controlled by the very corporations that stand to lose money if the environmental issues are addressed."

"Now you sound like the one who's paranoid," I said.

"Really? Did you know that 60% of the articles you read about in your daily papers dealing with social and political issues are taken from press releases issued by the corporations themselves or from special interest groups shilling for the corporations or politicians with environmentally friendly sounding names like: 'Council for Forest Conservation' or 'Friends of the Ocean'? They issue press releases citing studies—usually undertaken by their own paid scientists—showing that their products or actions are safe for the environment. Or, touting their million dollar contributions to worthy causes given from their *billions* of dollars in profits that were earned from their destructive business practices. It's called Green washing—the use of Public Relations to transform a company's image from one of being damaging to the environment to one of being eco-friendly."

"That really happens?" I asked.

"Michael, bless you, you are so naïve sometimes."

"That's just a nice way of saying I'm clueless, huh?"

"I'm afraid it is, yes. Do you realize that fewer than 20 corporations own over 90% of the major media outlets—that is, the radio and television stations, and the newspaper, magazine, and book publishers?"

"So, why do you think this book will be received any differently than any of the others? I will be ridiculed as just another nut case crying wolf—especially in light of the fact that I am claiming that Mother Nature is the co-author. If, in fact, we can get the book published at all. I mean, with the companies that stand to be threatened by what you are saying having control of the publishing industry."

"You are right. You will have those who will say you are just another alarmist or a nut case. You will have those who will say that you are just trying to make a quick buck by exploiting peoples fears because that is what they do. But, I have faith that it will be published and that enough rational, open minded people will read it and respond to it in a way that will overcome its detractors."

"Now who's being naïve?" I asked. "Even if we can swing getting it published, we already agreed that there are a lot of quality works already available that deal with this subject matter that are largely being ignored."

"True. But, there are also some that have had an impact as you know from your readings prior to being involved in this project and from the research you have done over the past few days. Books such as Silent Spring, Diet For a New America, The Last Hours of Ancient Sunlight, Ishmael."

"Yes, but those are exceptions to the rule. And, I don't think that I am even close to being in the same class as those writers. I mean they are messengers for goodness sakes."

"Well, whether you believe it or not, you are in their class."

"Oh, come on."

"If you don't believe me, ask them. I'm sure they would be the first to tell you that they don't consider themselves special or in another class from anyone else. I'm guessing that they just wrote from the heart and trusted the process."

"But, still...me?"

"Okay, if you can't accept that you are worthy of this, can you at least believe that Mother Nature is?"

"That I can do. But, I still think people are going to believe I'm nuts."

"That's out of your hands. Besides, we both know that there are a lot of people who think that already." She smiled sweetly.

"No there," I hesitated.

"Yes?"

"Well, maybe a couple."

"Maybe a couple?"

"That's only because."

She tilted her head, waiting.

"Oh, bite me." Weak I know, but again, all I had.

"Maybe we should get back to work," she said. A car horn sounded and I glanced towards the street.

"Yeah, maybe we should," I agreed. *Friday rush hour. I hope we finish before traffic gets too congested.*

"Patience, Michael, this is important." *Damned intuiting.*

"Yeah, yeah, yeah, carry on," I said.

"Very well," she said. "We were discussing how the great civilizations of the past 10,000 years have destroyed themselves. The means of their destruction is well documented and invariably follows the same pattern. I'll just give you the Readers Digest condensed version."

"I appreciate that."

"The civilizations in question achieve cultural success in a geographical area and their population begins to outgrow the resources at their disposal. In order to continue to grow and feed their expanding numbers, they have to resort to progressive deforestation and agricultural expansion, which they do by conquering the neighboring tribes and confiscating their land. The decreased vegetation cover changes the climate of the immediate area and causes sporadic rainfall cycles and wide spread soil erosion. Harvests decline in yield, which results in famine. So, more conquering and expansion. Eventually, this leads to growing social unrest between the 'haves' and the 'have-nots' that this type of culture invariably produces. The outgrowth of this is an ever-increasing reliance on religion and public ritual as the people, having falling into fear and despair, look for solace and salvation. That's why your archeologists find escalation in the construction of temples and tombs near the end of these civilizations. They weren't symbols of wealth and good times, they were a desperate attempt to appease the 'Gods' and take away the pain and fear of the people. You lose your connection to nature and you lose your sense of the divine. And, you lose track of who you really are. A conquering army is usually given the credit for the fall of a civilization, but usually those armies are just cleaning up the mess that the conquered made years, sometimes even centuries before their actual fall and that left them vulnerable to being conquered in the first place."

"It's kind of like saying a lifelong alcoholic died of liver failure," I said. "Cirrhosis of the liver may get the official credit, but it was the damage done to the system by the drinking that led up to it."

"An accurate and colorful analogy."

"Colorful, I like that."

"I'm so happy," Mona said.

"No need to get snippy."

"Sorry. This topic inspires snippiness."

"Then go ahead and snip away," I said. "Just know that it's an unattractive side of you."

"I can live with that."

"I figured as much. I just wanted to note it for the record." I had watched a lot of the Bush/Gore election proceedings with all its legalese.

"So noted," she said. "As I was saying, the hoarding lifestyle and the resulting empires continued merrily on their way through the centuries in direct opposition to the history up to that time."

"How so?" I asked.

"Many tribes and cultures in the past tried the agricultural lifestyle at one point in their history, but all of them eventually recognized the harmful consequences of that lifestyle and abandoned it to return to the tribal life. All except one, that is—yours."

"Hmmm." What else could I say?

Mona gave me a wry look and continued, "The events with the most impact as to today's problem, besides the Agricultural Revolution that we've already discussed were the Industrial Revolution of the early to late 1800's in Europe, then in the United States, and the discovery of oil in Pennsylvania in the mid 1800's. The assault on nature really escalated at that point with the mass introduction of chemicals. The air and water pollution that the planet suffers from today began to be noticed during this era and in 1863 the first air pollution regulation of the modern age was enacted in England."

"Almost 140 years ago," I said. "Wow."

"Yes. Over one hundred years and you humans are still emitting tons of toxins into the atmosphere every day. Doesn't take a brick wall to fall on you."

"Snippy."

"Yes, well."

"Wait, you said the first laws of the modern age. What did you mean by that?" I asked.

"Well, in 1273, also in England, they enacted a smoke abatement law that was basically ignored for years. But, in 1306 the owner of a manufacturing concern was charged with burning coal and this law was used to convict him."

"What was his fine?" I asked.

"He wasn't fined."

"Bought his way out of it, huh?" *Business as usual.*

"Not really. He was beheaded." *Then again.*

"Whoa, radical."

"But, effective. For awhile anyway. But, as technology advanced and fortunes were made, the split between the haves and the have-nots grew. Economics became the rationalization, justification, and motivation for every action. Cities with factories and rail yards became the norm. Man continued to alienate himself from the natural world. Nature truly became just another thing to conquer and subjugate to man's wants. Notice I said wants, not needs. That helped to solidify the proposition that man was here to rule the rest of nature and could make the life and death choices for nature: cows lived, wolves died; chickens lived, foxes died; hoarders lived, tribes died; rich lived, poor died."

"That's awfully harsh," I said.

"Maybe so, but it doesn't make it any less true."

"I don't think it's true that we just let the poor die."

"Really. Then explain the process to me."

"Uh, it's not that simple," I hedged.

"Yes, it is that simple. Even with the environment as damaged as it is today, you still have enough food production taking place and enough food stored away to feed everyone who is hungry without taking food out of the mouths of anyone else. Yet, 60 million people per year die of starvation worldwide—40 million of them children—due to the 'simple' fact that they are poor. You play it off as politics, but if you were truly a spiritually based nation 'under God' there wouldn't be any starving masses. There wouldn't be any homeless population except for those who chose to live that way. There wouldn't be any people dying for lack of health care. You wouldn't have had over 18 *million* environmental refugees since 1992 just because of manmade disasters such as deforestation. These issues are issues only because the people who are suffering from the issues are poor. It never ceases to amaze me that you self-proclaimed religious, righteous people who believe so strongly in God the Creator treat His creations with such disrespect."

"Okay, okay you win. We are selfish and cruel." I felt something crawling across my ankle and looked down to see an ant. I carefully brushed it into the grass. Mona smiled and nodded her head in approval.

"You are not selfish and cruel," she said. "Not really. You just act that way at times. For the most part you are just scared."

"Scared?"

"Yes. I mean, look at how you live. The division between the haves and have-nots has grown so large that your planet is consumed with fear manifesting itself in violence. You have locks on doors to keep people out and locks on doors to keep people in. You have fences to keep people out and fences to keep people in. Security systems and guards and weapons to keep people out or to keep people in. You must protect your possessions at all costs. You have conservative 'right to life' people who are against terminating an unwanted pregnancy even if it means the pregnant woman dies leaving the newborn and four other children motherless. And, a good many of these same 'pro lifers' support capital punishment. If that's not enough irony for you, it's these same conservatives who rail against government intrusion in their lives while calling for *more* laws against human behavior."

"Yes, the party of Lincoln has really gotten away from its roots," I said.

"Well, it isn't like the so-called liberals are any better. Their only answer to everything is 'more'. More welfare, more food stamps, more homeless shelters, more programs. Year after year after year. And, year after year after year the ranks of the disenfranchised continues to grow. There are more people living in poverty, more people starving, and more people homeless than at any other time in human history. Why? Because the programs aren't working! But, God forbid anything changes or that you try something different. No, just keep throwing more money and more resources into the same inadequate programs. What kind of sense does that make?"

"You really believe that those types of programs don't work?" I asked.

"Ask them," Mona answered, pointing to the homeless people.

"You can sure brighten up the day, you know? Jeez, I feel like wandering into traffic."

"Gosh, I'm sorry. Is this disturbing you?"

"Hell yes it is. Who wouldn't be disturbed by it?"

"You would be surprised. But let me continue to disturb. That's one of our goals—to disturb the comfortable."

"That ought to be fun," I grumbled.

"I know," she said, with a big smile. I was being facetious; she was looking forward to it.

"We were talking about man's alienation from nature," she continued. "And how that led to his justifying the subjugation of nature to the sole benefit of man. Perhaps the most tragic consequence of this shift in attitude was the loss of human dignity through slavery, a practice that became an acceptable way of life in many areas of your world. It took the bloodiest war in the history of you humans up until that time, your American Civil War, to bring about the end to slavery and that was only in the United States. It continued for another 120 years in South Africa and is still taking place in some areas of your world today. The saddest fact about slavery is that it's based on economic considerations. Human lives vs. profits. Money. That great motivating factor that we spoke of before."

"That is a sad comment on our society," I said.

"It's not only your society. It is the mindset of all countries that have cultures even remotely similar to yours."

"Really?"

"Yes, really. If you study your history books, at least the honest ones, you will find that economics has been the primary motivator in all of your major wars. In the case of the Sumerians,

Greeks, and Romans, the need for more and more land and resources was the motivator. The American Revolution was in large part a response to the British Empire's excessive taxation of the colonies and the corporations of the Empire engaging in price gouging and other monopolistic practices. The War of 1812 was fought over trading issues, as was the Spanish/American War of 1898. The Mexican/American War of the late 1850's was over the land grab of Texas.

"The United States was pulled into World War I because it was supplying its allies with arms, a strictly financial decision at the time. World War II initially started because France objected to the Customs Union that Germany wanted to enact with Austria. I have no doubt that Hitler would have found another excuse to do what he did, but that's beside the point. The United States managed to avoid being drawn into the conflict until the Japanese attack on Pearl Harbor, which was in large part a result of the U.S. Navy blockade of oil shipments to Japan. Korea and Vietnam were ostensibly about the capitalistic vs. the communistic ways of life—which one was the best way to accumulate wealth and power. The Gulf War was fought, of course, over oil, the loss of which would have caused an economic hardship for the industrialized countries, especially the United States. Ironically, none of these wars would have come about had the tribal lifestyle remained the dominant way to live."

"You really believe that?"

"Of course. Again, study history, the answers are there. Bottom line, war is big business. World War II especially was a huge boost to the American and world economies helping bring about the end to the Depression and bolstered by the rebuilding and subsequent financial success of Germany and Japan after the war. The terrible economic depression of the 1930's was replaced by the post-war boom and viewed with much gratitude. But, it also rapidly accelerated the destruction of the planet that we see today. Agriculture was undertaken by huge conglomerates and aided by the use of fossil fuel burning machines and pesticides. At that time, pesticides were thought to be such a wonderful aid to the human condition that Paul Muller won the Nobel Prize in 1948 for his invention of DDT, a pesticide that has since been banned because of its lethal effects."

"Talk about your irony. The Nobel Prize for a fatal chemical."

"You humans do love your irony. The post World War II industrial boom also brought about the mass migration to the suburbs that triggered the decay of the inner cities. The land necessary for this expansion came at the expense of small farmers. Forests were cut down to provide the space and the lumber necessary to meet the housing needs. The need arose for families to obtain a second automobile so mom would have transportation to take the kids here and there and for her daily shopping chores that lead to more pollution and more need for petroleum products. Again, heavy on the irony."

"I'm beginning to see a pattern here," I said,

"You haven't seen anything yet. The 1950's saw the first acid rain caused by this massive industrial buildup. The economic growth in the U.S., and worldwide with the incredible growth in Germany and Japan, continued through the 1960's and 1970's and was aided somewhat by the money that was poured into the various space programs to combat the Cold War threat at the time and into the military-industrial complex fueled by the Vietnam 'conflict'.

"Finally, air and water pollution began to be noticed and publicized by people like Rachel Carson, Edward Abbey, and Aldo Leopold. Your world didn't take these warnings seriously until the mid-70's and it wasn't until 1977 that your President Carter enacted tax incentives for alternative energy research. Unfortunately, your President Reagan did away with those incentives a scant four years later under pressure from the large corporations who had contributed heavily to his campaign fund. Environmental issues were not exactly at the top of the things-to-deal-with list in the 1980's, the infamous 'me' decade. The rich continued to get richer, the poor continued to get poorer, and the planet continued to get plundered."

"That was an incredibly narcissistic time," I said.

"It certainly was. Too bad nothing much has changed."

"Meaning?"

"Meaning, your culture is still extremely selfish and self-centered. In other words, very fear-based. Except for a brief period in the late 80's and early 90's when the self-help movement and recovery from your various forms of addiction was in vogue and it was considered cool to be in a 12-Step group and/or therapy, your narcissistic life view has continued unabated. People's desire to get help for their addictions and dysfunctions—that were ironically brought on as a result of your separatist, closed off, materialistic society in the first place—have been turned into jokes and one-liners for your entertainers and politicians."

"I don't think that's quite accurate," I said. "I know a lot of people who are into spirituality and personal growth. And, not because it's 'cool'. They truly want to grow and be better people."

"Of course you know those kinds of people. That's what you have been seeking for yourself so you have gravitated to and attracted people of like mind. But, overall, people such as yourself are a vast minority in your society. But, speaking strictly about the ecology movement, it began when small groups of environmentally concerned citizens began to form organizations such as The Sierra Club in the late 1800's. The ranks of these and other organizations have continued to grow as people have become more and more aware of how truly serious is the damage that has been done to the planet. Aware in large part because of the work of groups like those mentioned above and the writings that we cited a few pages back. Unfortunately, more than a few of the environmental organizations have been corrupted by money and taken over by the corporate mentality and have become nothing more than fronts for Big Business. But, more about that later.

"Adding to the problem is the fact that the majority of the population in the U.S. and the other industrially based countries around the globe are still in denial of the facts or just plain don't know what the facts are in the first place. People still tend to think the environmentalists are, as you said, 'alarmists'. That everything will just magically work out on its own. That science will miraculously make everything okay. That somebody else will take care of the problems. That it's all just leftist propaganda, another way to insinuate the government into people's lives. That the environmentally sound products on the market today are just the newest way to make money and aren't really necessary. In other words, most people are woefully uninformed, chronically cynical, and need to be convinced that the problems are real."

"I must admit that if most people are as uninformed as I am, then we are in deep doodoo."

"Again, colorfully put, Michael."

"Thanks," I said. "I know that most of my friends and family and the people I work with would fit in the same category."

"Sad, but true," she said. "And, things will only get worse unless this ignorance is confronted."

"That is not going to be easy."

"Obviously, Michael."

"Again with the snippiness."

"Yeah, yeah, yeah. As I mentioned briefly before, the gap between the haves and have-nots has continued to widen. Over the past eight years, despite unprecedented economic growth in your country, the difference between the average corporate executive's salary and the average working class individual's pay has jumped from 19 to 1 to 415 to 1."

"415 to 1," I exclaimed.

"Yep."

"So if I make $300.00 per week, the executive makes, let's see…about $120,000 per week."

"$124,500 per week to be exact," she said.

"Thanks." I gave her a nasty look.

"But, who's counting?" she said and smiled.

"That is unbelievable. And, grossly unfair."

"Unfair, yes. Unbelievable, not from what I've seen of your culture. Should I go on?"

"I don't know," I said. "I'm getting depressed."

"We are almost to the end of the history lesson."

"Thank God," I sighed.

"Then we can get down to the *really* disturbing stuff."

"Tell me you're kidding."

"Not kidding."

"Wonderful. You wouldn't happen to have any anti-depressants on you, would you?"

"No, sorry."

"Yeah, me, too."

"Anyway, cancer, AIDS, TB, and various other viral diseases have all seen an increase in incidence and some are even developing resistance to treatments faster than the scientists can come up with them," she said. I was too stunned to even comment at that point. On she went. "Voter apathy is at an all time high, just at the time when you really need grassroots involvement. Most people today in your country have a 'what's the use' attitude because your elections have turned into beauty contests, a battle of sound bites, and negative campaigning. Then, the election winner is so beholden to special interest groups, lobbyists, and big business because of campaign contributions that effective, meaningful legislation is almost impossible to achieve. It gets too watered down to do any good in an effort to keep the interest groups happy. And, if your most current presidential election is any indication, things don't look to improve much."

"Let's stay on topic, shall we?"

"As you wish. The decay of the inner cities has spilled over to the suburbs and rural communities as drug abuse, random violence, poverty, promiscuity, teen pregnancies, single parent homes, abortion, assisted suicide, lack of education, and moral degradation have become the norm. I suppose that's not surprising considering the behavior of your most recent ex-President. What a wonderful example he set.

"Basic respect for life has been compromised beyond belief. And, that's human life I am speaking of—it's no wonder that the extinction of up to 130 species of plants and animals daily fails to register. Or that trophy hunting has grown upwards of 70% over the last decade. Then, overseas you have genocide, famine, health plagues, slavery, political corruption, war, and economic collapse—all serving as crude and ineffective forms of population control."

"Why are you telling me all this? What's the point?" I asked.

"The point is that the problems you are facing today didn't just pop up all of a sudden. The results of the agricultural revolution of 10,000 years ago is an example of a phenomenon known as Sensitive Dependence on Initial Conditions: a complex, chaotic event doesn't necessarily come from a complex, chaotic stimulus; it most likely began with something simple, like farming. The point is that the thinning of the ozone layer was discovered three decades ago; global warming was predicted over a century ago; the first legislation to deal with air pollution is over 700 years old; water pollution has been a problem for 50 years; soil erosion and deforestation have been happening for millennia. The point is that species extinction directly attributable to the actions of humans has been going on since the 1500's when Dutch explorers who had landed on the island of Mauritius killed *every* tortoise on the island to replenish their food supply before setting sail again."

"That was a whole lot more than one point. But, who's counting?" I asked, smiling.

"Okay," she continued—not smiling. "Then let me try to narrow it down to one main point: you have failed to learn from history. And, as one of your old sayings goes: 'if you fail to learn from history, you are doomed to repeat it'. Is it your wish to become extinct?"

"Of course not. That's insane."

"So is doing the same thing over and over again and expecting different results."

"Okay, okay, I get it," I said. "Now, can we get this history lesson over with, please? My fingers are starting to cramp up; my head is about to explode. And, I want to go home."

"And you call me a whiner." I started to respond, but Mona held up her hand and smiled. "I only have one last little tidbit."

"Halleluiah."

"The Aborigines that we spoke of before?"

"Yes?"

"By the most conservative of estimates, they have lived the same tribal lifestyle for over 50,000 years. In fact, they are the oldest continually maintained culture on Earth—what's left of them anyway. And, in all that time, they have caused no pollution, endangered no species, destroyed no forests, and depleted no resources. Yet, their lifestyle has been sacrificed so those of the hoarder culture can have 'things', can feel a false sense of control. Ironically, this is when your species gave up any semblance of control and started living in fear. You quit trusting in the natural cycle of life, in a way of life that had succeeded for hundreds of thousands of years, and decided that your new way was better. But, all is not doom and gloom."

"You could have fooled me," I said.

"Really, it's not. Because of your population explosion, you have more great minds to work on these problems than at any time in the history of your species. You also have the genetic history of the tribal lifestyle in your DNA. You merely need to tap into it and initiate a change in consciousness that will begin to turn things around. That doesn't mean that you can return to the same tribal lifestyle that your ancestors lived. Six billion people, and counting, can't live that same way. But, you can use your collective experience to find a way that will work for you now and provide a clean environment for future generations. You will suffer some hardships, but you can avoid extinction—if you enact change. But, it has to start now. Mother Earth is getting very tired. She has been weighted down, gobbled up, used, and abused unmercifully over the past 10,000 years."

"Let's enact some change then," I said.

"Very well. Now that the history lesson is completed we can get down to the specifics of the environmental abuse and some solutions. But, I suggest we break here. You have been bombarded with some heavy stuff today and undoubtedly could use some time to absorb it."

"This is true."

"Any questions at this point?" she asked.

"No," I replied. "I am too overwhelmed right now." My brain was on autopilot and my body felt like it was made of lead.

"Then, I suggest you head home and get some rest," she said.

"Good idea," I agreed.

"I'll be in touch soon to schedule our next meeting."

"In touch how? You going to 'channel' the information to me?"

"Not to worry," she answered. *Easy for you to say.* She flashed me a sly smile that gave me the shivers. "Oh, and in the meantime you may also want to tell Lily what's going on."

"Then again, I may not want to…but I guess I better." I hadn't really put any thought into that dilemma. I mean, I had been a little preoccupied. I could foresee some problems. After all, I had agreed to help Lily with her book (or two) and now I was going to have to shift my efforts into writing a book of my own. But, maybe I could do both. Sharing the information from Mona would be no big deal. I could just make copies of it for Lily and she could use what she needed. But, telling her about where the information came from—a woman claiming to be Mother Nature, especially a beautiful woman—now that was really going to be fun. Mona stood up, preparing to depart.

"I'll leave you with your thoughts," she said, that sly smile still on her face.

"Yeah, thanks, Mona," I replied.

"Just be honest, it will be okay," she said as I busied myself with putting away my pen and pad.

"You don't know Lily," I said looking up to see what she would say to that, but she was gone. I jumped to my feet and looked behind the tree she had been leaning up against. That was the only place she could have disappeared to so quickly. But, she was no where to be seen. That woman was spooky. I slung my daypack over my shoulder and walked to my car.

INTERLUDE

INDECISION

I went home as Mona suggested, but as for getting some rest; well that was another matter. My mind was operating at the speed of a NASCAR driver on crack as it raced through the day's events: my realization that Mona is really Mother Nature; the history lesson that she had given me; the fact that I was going to have to come up with some way to explain all of this to Lily. And, I couldn't find the brakes. Despite my fatigue, I felt like I had just eaten a pound of chocolate and chased it with a double espresso.

After a while I was able to focus my mind on one topic—what to tell Lily. I ran through various scenarios. I could tell her that I had decided to write a book of my own as a side project. Technically, that was the truth. And, since she was writing one (or two) of her own, what could she say? Of course, I would have to leave out the part about working with Mona. The problem with that idea was I didn't know if Mona was going to want to eventually meet Lily or involve her in the project in some way. Or, when and where Mona would decide to meet again. So much for focusing on one topic.

So, if it came to it, how would I explain Mona? On the one hand, Lily was generally very open about happenings and relationships of a less that conventional nature. I mean, I am talking about a woman who has been involved in seances, who has dabbled in witchcraft and casting spells, and who claims to see auras. Not to mention her bouts of looking for portals to different dimensions, usually while sleepwalking. (I was introduced to that little idiosyncrasy when I found her banging around in my closet one night.) The point being, she is not someone who is a stranger to the unconventional. On the other hand, no matter how often she tells me that she is not a jealous person, every time I even mention another woman's name in passing she immediately wants to know the specifics of my relationship with said woman and what said woman looks like. So, how well she would handle my telling her that I was not going to be putting as much effort into her project (or two) because I was going to be doing a project of my own with another woman—again, a woman who was claiming to be Mother Nature and just happened to be drop-dead gorgeous. Well, I could see that testing the bounds of even Lily's self-proclaimed open mindedness.

After a few hours of running through these and other mind games and trying out a few mock conversations, I decided to deal with it by not dealing with it. I took the telephone off the hook and avoided checking my e-mail. I distracted myself by reading until I finally dozed off around midnight. Nikita, the Femme Fatale of the feline world awakened me at 5:00 am, even though it was a Saturday and I didn't have to work. Nikita hasn't grasped the concept of weekends. She figures since I'm up at 5:00 am five days in a row (Monday through Friday) then I should be up at 5:00 am on the sixth day, too, and the seventh and so on. So, when I opened my eyes, there she was, her nose inches from mine, meowing in that way of hers that I have come to interpret as 'Hey, I've been up all night, by myself, going from windowsill to patio door and back again and I'm bored out of my skull. Therefore, it's time for you to get up and entertain me'.

I surrendered to the inevitable and got up. I played with Nikita while waiting for the coffee to brew. Her favorite toy is a plastic clothes pin (go figure) tied to a string that I toss around the apartment for her to chase. By the time the coffee is ready, she has usually had enough of this and goes off by herself. I think she just knows by now that I am up to stay and she doesn't have to keep me busy. She can go about her business knowing that I am there if she has a need to be met that she can't fulfill on her own. Usually her business is to sneak off to the bedroom and curl up on the recently vacated bed for—dare I say it—a catnap. The first few times this happened, I would wake her up by going into the bedroom and saying 'You wake me up, then 20 minutes later go to bed yourself!' This was routinely met with a weak 'meow' followed by her rolling over and going back to sleep. She's lucky she's so damned cute.

I had acquired her about six months previously after someone in the apartment building I used to live in had abandoned her (I just moved into my present location three months ago). She was hanging around the door one morning when I went to work and was still there when I came home in the afternoon. She would try to follow me into my apartment. Outside of a few pats on her head, I managed to ignore her for three days. One of the other tenants had put food and water bowls in the hallway for her. I figured whomever had lost her would find her soon. Wrong. I ended up asking around the other apartments and nobody claimed her. Personally, I think the punk drug dealer upstairs had been the one to abandon her, but I could never prove it.

I drove through the parking after work the fourth day and saw that she was in her usual spot by the door, so I went to the store and got a kitty starter kit—cat litter and box, food and food dish, flea collar and flea dip. When I returned home I let her follow me into the apartment. Luckily, she knew the intended use of the litter box. The next day I took her to the veterinarian, had her checked out, and got her shots and, the rest, as they say, is history. I had not wanted to get another pet. I had to have my dog of thirteen years put down two years prior and after getting over that had begun to enjoy the freedom of being able to come and go as I please without having to make arrangements for a pet.

But, Nikita was just too damned friendly and pitiful to ignore. I just couldn't not do something. I figured that she had been dropped into my life for a reason. (Seems like a pattern with the females in my life lately. I had met Lily online under freakish circumstances at a time when I had no intention of entering into a relationship and there was this thing with Mona. Wonder what this Karma is about?)

Anyway, after Nikita had gone back to bed, I drank my coffee while reading the newspaper and watching CNN. I then went for my morning run. It was a bit earlier than usual, but I use my run as a time to meditate and I wanted to avail myself of that in a last attempt at deciding on what course of action I was going to take with Lily. Fortunately, this was one of the weekends that Lily had a program and would be busy most of today and tomorrow.

I set off on the five-kilometer course that winds through my neighborhood. It was a lovely Mid-western spring morning: brisk, but sunny and around 55°. I had on shorts and a cutoff sweatshirt. It was a bit cool at the start of the run, but I knew that I would be plenty warm by the time I finished the first mile.

About halfway through the course the route takes me past a small park with a few playground staples: swing set, jungle gym, seesaw, and a basketball court. As I passed the park I noticed someone hop off one of the swings and jog towards me. To my surprise it was Mona, but in the African guise and sporting a bright red jogging outfit with matching shoes. I slowed down abruptly and almost tripped over my own feet. She fell in along side me and said, "Don't stop." So, I regained my balance and resumed my normal pace. *That change in appearance weirds me out.*

"Sorry," she said, intuiting my thoughts.

"No problem," I said. "So what brings you here?"

"I just wanted to make sure that we set a time for our next meeting," she said, "before you talked to Lily." I didn't bother to ask how she knew that I had yet to talk to Lily.

"And, when would that be," I asked, referring to a meeting time.

"How about at 11:00 this morning at Audubon State Park? I thought we could go for a hike."

"That sounds good to me. It's supposed to be a beautiful day." Then, remembering whom I was talking to, I had to laugh at reporting on the weather. She was gracious enough to not mention it.

"The parking lot by the camping office okay with you?" she asked.

"Sure."

"Great, see you then." By that time we had just about circled the park. A small grove of trees separated it from the residential homes in the area. Mona veered off into the trees and, just like at the library, disappeared.

"Yeah, see you then," I said to the air and finished my run. I decided to tell Lily the only thing I could...nothing. I thought it best to wait a few days and see how things went with Mona. Yes, I know, big mistake.

When I got home, I made another mistake—I answered my telephone. I had put it back on the hook before I left for my run assuming Lily wouldn't be calling since she was at work. Wrong, again. She was at work, but on a break.

"Why was your line busy all night last night?" was how she began the conversation.

"Nikita must have knocked it off the hook," I said. "I didn't notice it until this morning." *Wow, I came up with that pretty fast. Maybe not being exactly forth coming (okay, okay—lying) with her wasn't going to be as hard as I thought.*

"I can see that crazy cat doing that." *Plus, she bought it!* "So, did you have a chance to read the stuff I faxed you?" she asked.

"Yeah, I did," I replied. "You're right, his writing is very technical."

"It is, isn't it? Now you can see why he, or at least his publisher, wants someone to lighten it up."

"Yeah, really. I was able to get some research done, too. I'm going to do some more this afternoon." *Well, that was true, wasn't it?*

"Really? It's supposed to be a beautiful day. You sure you want to spend the day cooped up in the library?" *Great, let's start with the guilt already, shall we.*

"Well, I feel kind of bad with you having to be inside all day with your program and all." Well, I was feeling bad. Hey, I can rationalize with the best of them. I also felt trapped, now. Past the point of no return. Too late to retract and tell her the truth.

"That's sweet," she said, intensifying my guilty feelings. We wrapped up our conversation after agreeing to get together for a late dinner after she got off work.

CHAPTER 4

RAINFORESTS

I arrived at the park a few minutes before eleven and found Mona sitting in the Lotus position on top of a picnic table at the edge of the parking lot. She was still African in appearance, but now attired in shorts, a chamois shirt covering a halter-top, and hiking boots of a strange synthetic material turquoise in color. She was in profile and looked radiant in the morning sunlight. I sat in my car not wishing to disturb her. There were a few other cars in the lot, but no one else was in sight. After a few moments, she uncurled herself and slid gracefully off of the table. I exited my Jeep. Turning to me, she smiled and said "Isn't it a beautiful morning, Michael?"

I smiled and nodded in agreement. It was indeed a beautiful morning. High wispy clouds floated across a robin's egg blue sky. The air was crisp and clean and fragrant with the scent of the pine forest I knew to be a few minutes walk from where we stood. The temperature had already risen to about 65 degrees and promised to get much warmer as the day progressed. Unusually warm for this time of year, but I wasn't complaining. Sounds of birds chirping and insects humming played rhythmically in the background.

I retrieved my daypack from the back of the Jeep. I had a few Energy Bars and two bottles of water along with the stuff I usually kept in my pack: a small first aid kit, Swiss Army knife, bandana, suntan lotion, and water proof matches. I had also thrown in a pen and notepad and my micro-cassette tape recorder. I couldn't help but notice that Mona was empty handed.

"You ready?" she asked.

"Uh, yeah, okay." I stammered.

"What?"

"Don't you have any water or anything?"

She laughed that musical, magical laugh of hers and said, "You are just too cute for words sometimes."

"Gee, thanks. What, you don't eat or drink?"

"Of course I do silly. I had something earlier. And, if I get hungry or thirsty on the trail I'm confident that I will find what I need to survive. And, you will, too as far as that goes."

"Dopey me. But, I think I'll bring my pack along anyway, if you don't mind."

"As you wish." She said this in the same voice one would use who is showing indulgence to a stubborn child. *I don't care. I'm not setting off on a hike, having no idea how long I'm going to be on the trail, without my provisions. Even if I am hiking with Mother Nature herself.*

"Oh ye of little faith," she said.

"No offense intended. Just being responsible," I said in defense.

"Nothing wrong with that. Now if we could only get you and your kind to apply the same principle to the environment."

"Getting snippy already, are we?"

"Just making an observation."

"Right." *This has the makings of a long day.*

I locked the car, stowed my keys in a zipper pocket, and slid the pack on my back. I hooked my tape recorder to one of the straps and turned it on. Mona started off in the direction of the pine forest.

"Did you get any rest?" she asked.

I snorted in reply.

"What? You had trouble sleeping?" She was acting so innocent. The key word being 'acting'.

"Of course I had trouble sleeping. My mind was on hyperdrive."

"I guess that yesterday was a bit of a shock to the system, huh?"

"A bit of shock to the system. Well, let's look at that. I find out that the planet that I live on is about to relegate my species to the category of 'extinct'; that I have been drafted to write a book on a topic that I am only vaguely familiar with—an undertaking that I feel vastly under qualified for, I might add; and, that I am to write this book with the help of none other than Mother Nature herself. Oh, yeah, then there is the matter of Lily. Shock to the system comes maybe within shouting distance of covering it, I guess."

"My husband will be helping us, too, don't forget."

"Oh, that's right. I also have to deal with you channeling the Darth Vader of Ecology."

"I guess this would be as good a time as any to mention that some of my crew will also be, uh, dropping in on us as needed."

"Hey, the more the merrier. Will they be here in person or will you be channeling them, too?"

"In person."

"Wonderful."

"I'm glad you are so amenable to it." She was trying hard not to laugh.

"Yeah, no problem."

We walked in silence for a few minutes.

"So you failed to tell Lily the truth." Mona said this as a statement, not a question.

"I didn't know how to broach the subject," I answered. "Without sounding like I needed my medication level checked. I decided to just let it slide for now."

"The longer you put it off, the harder it's going to be—for you to tell her and for her to believe you once you do."

"How about we just limit our discussion to the environment, huh?" *I'm not in the mood to receive relationship advice from a woman whose own engagement lasted four billion years.*

"Okay. I mean since your track record in relationships is so good and all."

"Mona."

"I'm just saying."

"Well, just don't say, okay."

"Okay," she said. It was quite obvious by her tone of voice that she had plenty more to say on the subject, but was going to keep it to herself…for now.

We again fell into an uneasy silence. Mercifully, the pine forest came into view.

"Let's go in and grab a log," she said.

As we entered the pine forest we were instantly enveloped in shade. The sun could only come through the dense foliage in scattered rays. Because of this, it was about ten degrees cooler under the canopy. Fallen pine needles provided a thick carpet to walk across. We soon came upon a clearing. There were small groupings of logs and old tree stumps to sit on, scattered here and there that had accumulated over the years, due to the efforts of previous hikers. We each picked a log and sat down a few feet from each other. "I thought we would start today's session with a discussion of the world's rainforests," Mona began.

"The setting is apropos, I guess. Even if this is not a rainforest. And, it's not raining."

"Now who's getting snippy?"

"I'm just saying," I said, mimicking Mona.

"Me thinks your snippiness stems from your guilt over lying to Lily."

"You can be highly annoying you know that?"

"Yes," she answered in a way that conveyed that she did know it…and didn't care.

"You were going to tell me about the rainforest," I said, hoping to move along before we really made each other angry.

"Yes, I was. Do you know what a rainforest is, by the way?" she asked. But, before I could give her my answer (which would have been 'not exactly') she held up her hand to stop me. "Never mind. I don't believe that a Q & A format is the best way for us to proceed at this point. I will just give you the information and you can stop me as you have questions."

"Fine," I grumbled.

"Fine." *This is going well.* Mona took a deep breath, let it out slowly, and then continued, "Technically there are three main types of rainforests throughout the world. They are mainly located in Central and South America, Southeast Asia, Australia, and Africa. Then you have your temperate rainforests in the Pacific Northwest here in the U.S. For our purposes I will treat them as if there were only one."

"Why?" I asked.

"Because they basically serve the same functions. The main one is to act as the lungs of the planet. A rainforest is made up of different layers. There is the emergent layer that consists of the tallest, oldest trees. Below that is the canopy that is a sort of interlocking of the crowns of the trees. Incidentally, this is where fully 2/3 of all species in the rainforest live. Below the canopy is the layer known as the understory which is a dark, wet layer made up of shrubs, vines, bushes, young trees, and such. This is the layer that is commonly known as the jungle. And, at the base is the forest floor: seedlings, small animals, soil, and such. Your scientists have estimated that the rain-forests are home to four million species of plant and animal including: 50,000 trees, 2,000 birds, over ¼ of your world's amphibians, and a few million insects."

"Estimated by our scientists? Don't you know for sure how many there are?"

"No. I haven't kept track of them all. I was there at the start, but evolution and extinction events have created havoc with census taking."

"All righty then. You mentioned that the trees are the lungs of the planet. Would you care to elaborate?" I asked.

"Certainly. A tree, through it's surface area—the area covered by it's leaves—uses sunlight as an energy source to convert excess carbon dioxide in the atmosphere into oxygen through a process known as photosynthesis. Humans, along with most other mammals and animals, breath in this oxygen. They, in turn, process the oxygen and exhale it as biological waste in the form of carbon dioxide. This is a finely balanced, integrated, cir-cular system. If you have spent more than a minute or so under water you know how important it is to have oxygen to breath."

"So, the rainforests are important. I get it."

"That is not the only function of the rainforests, though."

"What else do they do?"

"Well, rainforests—all forests for that matter—also function as aquifers. The leaf surface of one large tree can process moisture equivalent to a forty-acre lake. At deep levels in the earth, water has acquired high concentrations of minerals, particularly salt. Most of the minerals that are beneficial to the flora lie above the layer containing the salt. Groundwater, which is taken from above the salt water level, is drawn through a tree's root system. The trees use the minerals in the groundwater as nourishment, then release the excess moisture into the atmosphere as water vapor. In turn, this water vapor is condensed into clouds and eventually falls back to the earth as rain to strengthen the soil and provide sustenance to the other plants and animals—one of which are you humans. When the process is interrupted, salination occurs."

"What's salination?" I asked.

"When the water is removed from the soil it creates a downward draw into the soil and provides a place for the fresh rainwater to soak. This keeps the soil robust. When forests are cut down, the saltwater creeps upward to within a few yards of the surface where it begins to negatively affect the immune systems of the remaining trees, resulting in insect infestations and fungal infections. When the saltwater gets to within a foot or so of the surface, the soil becomes incapable of sustaining vegetation and the land becomes barren. Eventually, desertification occurs: the land becomes a desert. Barren land means no crops. No crops, no food. No food, well, I'm sure I don't have to spell it out for you."

"Uh, no, you don't. If we destroy the rainforests, we will get hungry and won't be able to breathe. Wonderful," I said.

"Oh, but wait, there's more."

"Somehow I figured that."

"This process also renders the water supply unsafe to drink because of the increase in the mineral content. So, you won't have any water to drink to wash down the food you won't have to eat to give you energy to breathe the air that won't be breathable."

I started to rummage through my daypack.

"What are you looking for?" Mona asked.

"My Swiss Army knife. I just hope it's sharp enough to slit my wrists."

"Wait, it gets worse."

"You would make a great crisis counselor on a suicide hotline, Mona."

"Thank you. As I was saying."

"I was being facetious."

"And, I was ignoring you." She flashed her sweet smile.

"Oh." I flashed mine back at her. Neither one of us was very convincing.

"Even though the rainforests only cover 7% of the earth's surface," Mona continued. "They are home to 50% of its species of plant and animal. The importance of this is that only 1% of the plants has been thoroughly studied for medicinal purposes. Yet that little bit has already yielded cures and/or treatments for such maladies as migraine headaches, cystic fibrosis, various forms of cancer and leukemia, cholera, malaria, ulcers, hypertension, syphilis, and AIDS."

"That's unbelievable."

"Want to hear something even more unbelievable?"

"I'm not sure, to tell you the truth."

"This is a good one."

"Coming from you, that doesn't ease my fears all that much. But, go ahead."

"Okay. One twenty-five acre tract of land in Borneo is home to 700 different species of trees. That's equivalent in number to *all* of the species found in North America." She smiled like a school child that had just given the correct answer in front of the whole class.

"That *is* more unbelievable."

"Want to hear another one?"

"Sure." I mean, how could I not. She was positively giddy. I hadn't the heart to dampen her enthusiasm. And, I was glad to see the mood lighten up.

"Cool. Remember when I told you earlier that everything had been provided for all the species to thrive?"

"Yes, I remember," I said.

"Well, there are approximately 125,000 species of plants in the rainforests. Just one plant, the rosy periwinkle, with only two of it's sixty alkaloids, has produced Vincristine and Vinblatine—drugs that are used to cure patients with acute lymphatic leukemia and Hodgkin's Disease—and between them generate over $200 million per year in sales. Tell me that's not the best of both worlds from your cultural viewpoint. Large profits and a cure for cancer, too!"

"And, you call me cynical."

"You are cynical."

"Well, so are you."

"No, I'm a realist."

"What's the difference?" I asked.

"Me, a realist, sees things the way they are. You, a cynic, see things in a negative light no matter how they really are."

"Oh. Okay."

"What, no argument, no debate, no denial?"

"What's the use? You would just twist things around until I was just too confused to argue anyway."

"Keep thinking like that and maybe someday, you too, can be a realist."

"If only I were bold enough to have such dreams."

At this point Mona paused and looked around the pine forest. She appeared to become restless. All of a sudden she said, "What say we stop the lecture for a while and take a walk?" Without waiting for my response, she hopped up and set off at full stride across the clearing. I scrambled to my feet, threw my pack over my shoulder, and ran to catch up.

We walked through the pine forest and back to the trail. We followed the trail for about ten minutes before coming to a fork. The trail to the left led downward to the river that ran through the park and eons ago had formed the gorge for which the park was famous. The trail to the right went up and ran along the upper edge of the gorge. Mona chose the right fork without hesitation. She obviously had a particular destination in mind. We hiked for quite a ways, making small talk in the interim, before stopping at a rocky shelf that over looked the river below.

Mona dropped into a cross-legged sitting position commonly known as 'Indian-style'. More accurately, so fluid was her movement, it was if she were a smooth wine being poured from a chalice. I noticed that despite our strenuous hike she was breathing quite normally. I, on the other hand, was panting as if I had just staggered across the finish line at the end of a marathon. And, I am in decent shape. Or so I thought. What a hit to my ego.

I took my pack from my shoulder and lowered it to the ground. I dropped beside it and extracted one of the water bottles from its pocket. I offered it to Mona. She smiled and shook her head 'no'. *Of course not.* I took a long pull then put the water back in the pocket. As I wiped sweat from my brow and neck, I noticed that Mona wasn't even damp.

"You seem a bit winded," she said.

"It's this fresh air," I replied. "I'm used to the air I breath having chunks in it."

"That's closer to the truth than you know," she said. "This may seem like fresh air to you, but believe me, it's far from fresh."

"Mona, compared to the air in the city, this is pristine."

"True, compared to that, this air is pristine. But, unfortunately, the air pollution generated in the city doesn't stay in the city."

"But, we are twenty-five miles from the city. You mean to tell me the pollution makes it clear out here?" I asked.

"Twenty-five miles is a stones throw. Global air currents are capable of transporting carcinogens and toxins half way around the world."

"Get out of here!"

"I'm serious."

"Half way around the world?"

"Yep."

"How?"

"We will get into that when we discuss air pollution. For now let's stick to the rainforest."

"As you wish."

I was also hungry and pulled both of my Energy Bars from my pack. I offered one to Mona, but she refused this, too. *She must have had a hell of a breakfast.*

"As I mentioned earlier," Mona said. "There are over one hundred species of plant and animal becoming extinct daily. What took billions of years to achieve is being destroyed, in some cases, in a matter of days."

I concentrated on my energy bar.

"On average," she continued. "Up to 8,000 acres of rainforest are being destroyed daily. So, along with destroying your food, water, and air, you are destroying the possible cures for your most deadly diseases."

"Unreal."

"Yes, it is…especially for such a supposedly intelligent species." She stopped, raised her eyes to the sky, and cocked her head to one side. *Uh oh, I sense another channeling coming on.* "My husband wishes to say something."

"Oh, goody."

"There is also an emotional and psychological tie to the trees," boomed Lean Gene's voice out of Mona's mouth. Even though I had witnessed this once before, it still blew my mind. Kind of like watching Jim Nabors sing or Mike Tyson talk. "Their many sizes and shapes have provided children of all ages with branches to climb and limbs to support tree houses. Their bark has provided romantics a place to carve their declarations of everlasting love. They sprout leaves in the spring to herald the start of a new season of growth. These same leaves provide a shady reprieve from the hot summer sun. And, in the fall they put on a show of colors seen no where else in nature. They have provided wood to shelter you from the elements. They have given their lives to provide the paper on which this will be printed. It is hard to find a nobler example of creation. Yet, you insist on destroying them en masse. It must stop. At present rate, the rainforests of North America will be gone by the year 2030. By 2050, all of the rainforests of the world will be but a memory. Not a rosy future for your children, is it?"

I wasn't sure whom I was answering, but I looked at Mona and said, "No, it's not."

Mona shook her head and said, "I'm back."

"Good," I said. "I have to tell you that is very disturbing and I don't mean in a good way. I mean in a 'call Mulder and Scully' kind of way."

"I've told you before, Michael, he is harmless."

"I know what you've told me, and I have to admit I was surprised at how soft he sounded talking about the beauty of trees, but still."

"I see your point," she said, laughing. "But I can't guarantee that it won't happen again."

"I'll deal with it."

"I'm sure you will. I have something else I need you to deal with now, though."

"This ought to be good," I mumbled.

"To paraphrase one of your politicians," she said with obvious distaste. "It depends on your definition of 'good'."

"Cute," I said.

"Thank you. Anyway, when you read or hear about the decimation of the rainforests or the environment in general, it seems that all of the statistics and emphasis is on how it is contributing to pollution or global warming or how many plants and animals are being made extinct."

"So, isn't that the point?" I asked.

"Yes, but there is one fact that is usually omitted."

"Against my better judgment, may I ask what that might be?"

"Absolutely. It is the fact that whole tribes of people are being uprooted and/or destroyed by this assault on nature. Remember the tribal lifestyle we spoke of previously?"

"Of course."

"Okay, well, all of it is in danger of becoming extinct. There are approximately 2,000 tribes worldwide, mostly residing in the rainforests of South America, Africa, Southeast Asia, and Australia. Besides the ones that we have already spoken of there are also the Pygmies of Zaire, the Yanomani of Northern Brazil, the Kuna of Panama, the Maya of Central America, the Yani of Indonesia, and the Penan of Malaysia. They are constantly being forced from their ancestral homes, their food supplies destroyed, and their very way of life threatened."

"Why is that happening? What is causing it?" I asked.

"You mean besides the well meaning, but arrogant missionaries from various religions who try to persuade the tribal savages that the life they have been living for thousands and thousands of years is wrong? That the so-called savages need to convert to whatever beliefs the missionaries believe in because those beliefs are the only right beliefs to have?"

"Uh, yeah, besides that."

"Well, loggers and developers cut down the rainforests to provide the housing industry of the Northern Hemisphere with lumber. That region of the world has 25% of the world's population, yet uses 80% of its wood. The governments of the countries—in collusion with businesses of the foreign interests—cut the trees down to provide agricultural space for crops. This is an abomination on so many levels.

"First, because of its specialized ecology, the soil of the rainforests is not rich in the nutrients necessary for this type of land use. Therefore, after a mere two or three crop seasons, the land becomes sterile. Second, when the trees are burned, large amounts of carbon dioxide is released into the atmosphere. Currently, 10% of the carbon dioxide being added to the world's atmosphere each year comes from this practice. If you remember what I said earlier about the functions of the trees, this means that you are burning the very mechanisms needed to process the carbon dioxide out of the atmosphere. And, again going back to the functions that trees perform, moisture is being removed from the atmosphere because of less leaf area to perform the evapotranspiration cycle, which leads to erratic drought and flood patterns, polluted rivers, and less rainfall.

"Those same governments also allow the forests to be cleared to provide pastureland for cattle to feed the carnivorous populations of the industrial countries. In fact, seven times more forests are cleared to provide pastureland for cattle than for all other development. And, usually the people made to work these lands are so poorly paid that they can't even afford to eat the meat that is raised there.

"And, the tribes that are forced to relocate add to the problem because they have to cut down more of the forests for new land space, to replace that which was taken, in order to plant new crops and build new homes. That is if they don't die first from diarrhea or dysentery caused by the mercury poisoning of the water supply. Or from having no immunity to diseases that are brought by foreigners."

"Why isn't anything done?" I asked.

"Money, of course. The corporations behind the destruction are allowed to thrive because of bad governmental policies in the form of subsidies, low taxation, bribes, and stolen property rights. Since 1945 alone, half of the world's rainforests have been destroyed due to actions such as these. In the 1990's, the average rate of destruction was 80,000 square miles per year that added 3 million metric tons of carbon dioxide to the air each year.

"The relocation of tribes and the encroachment of developers also exacerbate what is known as the Edge Effect. This means that species on the boundaries between the forests and the grass-lands are more susceptible to harm due to varying climates over a short distance and incursion by non-forest animals and humans. This can effect biota for up to one-half mile into the for-est and accelerate extinction rates. Normal, or what's sometimes referred to as background, species extinction occurs at the rate of one species every four years. The current rate is 1,000 times higher than the norm. This qualifies as mass extinction.

"It also creates a negative domino effect in that some of the species being killed off are what are known as keystone species. Keystone species are integral to an area because when they die, species dependent on them for their survival also die and on and on down the line. As the food supply dwindles because of these occurrences, the domino effect accelerates. When the human species invades the area the effect grows even more alarmingly.

Humans bring with them new creatures: dogs, cats, goats, and pigs, and worst of all, rats. Because rats feed on eggs, young birds, and reptiles it sets off another chain of destruction early in the life cycle. And, you have humans hunting, often times just for trophies, which obviously results in more killing of species."

"You would think," I said. "That with all our brainpower we would be able to figure this out."

"Ironically," Mona responded. "The evolution of the human brain has played a part in bringing about the current dire circumstances. When the climate changed two million years ago from one of your Ice Ages to one of a warmer climate, it forced Homo habilius from the forests to an area of open grasslands and triggered the evolution of Homo erectus. Homo habilius had been primarily vegetarian obtaining 80% of their diet from plants. Due to the changes in habitat brought about by the changes in climate, Homo erectus began to hunt more frequently in order to supplement his diet. This change in lifestyle fueled the evolution of the brain because Homo erectus had to be able to 'learn' to hunt and to remember seasonal changes and that beget a more complex life style. The hunter/gatherer was born. To sustain this life style, communication had to develop so they could work in teams. Eventually this newer, smarter brain figured out how to grow crops and herd animals and the precursors of your modern communities were born."

"I can't believe that cutting down trees has such far reaching effects," I said.

"That's the problem—you humans can't, or refuse, to believe it. But, you better start believing it before there are no trees left to cut down."

"Aren't we were already doing that? I mean, I have read about efforts and projects to plant trees to help make up for the ones being cut down."

"That's true and it's a start. But, obviously, it takes many years for a sapling to grow into a mature tree. If you continue at your current rate of destruction you will run out of trees long before the newly planted ones are full-grown. If in fact they are able to survive to maturity what with all the residual problems that we have just discussed such as air and water pollution and soil degradation. And, the lumber companies do a great deal of the tree plantings. The problem with that is they plant one species of tree for quick growth instead of a diversity of species. This practice fails to rejuvenate the ecosystem that has been destroyed by the slash and burn tactics in the first place."

"What can we do now to stem the tide?" I asked.

"Quit cutting down the old growth trees." *Well duh.*

"I assumed that," I said. "I meant, what could we replace them with? What should we use instead of wood?"

"We will get into the solutions later. There are still too many issues that we need to discuss that are tied to the problems of the rainforests. And, as such, great many of the solutions are tied together, too."

"It's your call."

"Well, thank you for your cooperation, Michael."

"You are welcome, Mona. So, what's next?"

"Well, the first issue we discussed when we began talking about the rainforests was how they are the lungs of your planet. So, I would next like to address the problem of air pollution."

"Okay, air pollution it is," I said as I shifted my position to get more comfortable. It seemed as if we would be there a while.

"I hate to stop when you are in such an agreeable mood, but I think it would be best if we saved that for another day. It's getting late and you don't want to keep Lily waiting." I guess we won't be here a while.

I glanced at my watch and couldn't believe that it was 4:30 in the afternoon. I mumbled something about time flying.

"How are you going to explain your sunburn from today to Lily? Strong fluorescent lights at the library?" I felt my forehead and it was indeed tender to the touch.

"I'll handle it." I still didn't want to talk with her about Lily.

"Okay," she said in that same tone indicating a wish to say much more about the subject. But, again, she opted to refrain herself. "Let's get you back to your vehicle then, shall we."

"That would be great."

Mona knew a short cut (of course) and we were back at my Jeep in no time. I stowed my pack in the back.

"Let's meet tomorrow morning at 9:00 am at Carillon Park, okay?" Mona proposed.

"The park next to the Power Company to discuss air pollution. Not a coincidence I take it?"

"Not a coincidence."

"Didn't think so. You need a ride?" I felt foolish asking her this, but it seemed the polite thing to do.

"No, thanks," she said.

"You have some sort of a 'beam me up, Scotty' thing going?" I asked. "I mean, the way you disappear and everything?" I climbed into the driver's seat.

"Something like that, but without the sound effects or the glittery particles," she responded.

"Cool. Can I watch?"

"There's nothing to watch. I'll see you in the morning. Tell Lily I said 'hi'." She began to walk into the trees.

"Very funny," I yelled at her back.

"Come clean with her, Michael," she said over her shoulder.

"Maybe I will," I said. But, she was gone. It wasn't like I could see her disappear, like she faded away or anything. She was just gone, like the trees and foliage had swallowed her up. She was right; there was nothing to watch. I started the Jeep and drove home.

INTERLUDE

DECISION MADE

Lily drove to my house when she got off work and we went to dinner at one of our favorite restaurants near where I live. It is an eclectic place where you can dress casually and get a quality vegetarian meal—we both favor that type of cuisine. She seemed a bit distant on the ride over, but I wrote it off to her being preoccupied with work. She was usually like this on the weekends that she had a program. I was hoping that she was so preoccupied that she wouldn't notice my sunburn. No such luck.

"I see you decided to forego doing research after all," she said as soon as we were seated.

"Not exactly."

"Not exactly? What, you doing commercials for Hertz now?"

I laughed at her joke and said, "I took my research to Audubon. Thought I would kill two birds with one stone, you know." I almost expected Mona to beam in and give me grief for using such a phrase.

"Good idea," Lily said. "Did you get much done or did you spend your time hiking?" She smiled in such a way as to let me know that if I had blown off doing research she wouldn't have blamed me. Of course, this just added to my guilt. But, at least she was asking questions, so far, in such a way that I didn't have to tell blatant lies. I could get away with merely telling Clinton-esque type lies.

"I was able to do some of both," I answered. "How is your program going?" Obvious attempt at changing the subject, I know, but I was reaching.

"It's going well. I have a mellow group and we're laughing a lot, which is always a plus."

"That's great. How many showed up?"

"Twenty-eight."

"Good numbers," I said. *So far, so good. Let's stay on track here.* Our waiter came and took our drink order.

"So, did you find out anything about Sustainable Development?" Lily asked. *Damn. There goes his tip.*

"Not a whole lot," I answered.

"Really? The library didn't have much on that subject?"

"I don't know, I kind of got sidetracked doing research on the rainforests." *Here we go.*

"The rainforests?"

"Yeah," I said with a weak smile. "So, are you sending many folks to treatment this program?"

"Five or six. What's the research on the rainforests about?"

"It's very interesting," I said. "Did you know that, among other things, the rainforests serve as the lungs of the planet?"

"No…I mean, yes I knew that, but that's not what I meant. Let me rephrase my question. Why are you doing research on the rainforests?"

"Uh, well, I just found it really cool." *Don't panic, think.* "And, a lot of the stuff I found out pertains to Sustainable Development. I'm sure that you will be able to use some of it in both of your books." *Not bad.*

"How does it pertain to Sustainable Development?" *Damn, she is persistent.*

"It's hard for me to explain without the material here in front of me. Let me get it organized and it will make more sense. It's still new to me, too."

"Okay," I could tell by the way she drew the word out—you know, okaaaaaaay—that she was getting suspicious. Luckily, the waiter returned with our drinks. Iced tea for me, White wine for Lily. He just may have redeemed himself.

For the rest of the dinner I managed to steer our conversation toward books and politics and spirituality. Afterwards we took in a movie, then returned to my apartment to retire for the night. It was a pleasant evening, but I felt an undercurrent of tension radiating between us. Gee, I wonder why?

We were up early the next morning thanks to Nikita. But, Lily had to get to work anyway. Over breakfast she asked about my plans for the day. I told her that I was going to treat myself to a day of leisure—a bike ride, reading a novel I had checked out of the library, perhaps a nap on the porch-nothing to do with work. *That isn't far from the truth. Well, not that far.* I had come to the conclusion that I could do Lily's research during the week on the internet while I was at work with maybe one evening thrown in at the library. That would leave the weekends that Lily had programs and maybe a night during the week free to spend with Mona. That was the plan anyway. Lily and I rarely spent time together during the workweek because I also worked two nights at a bookstore and liked a few nights free for myself. I just hoped that I could keep my double life from Lily until I could figure out the best way to handle the mess I had made by not telling her the truth up front.

After breakfast, Lily left for work. She said she would call me when she got home and let me know if she was up to getting together. These long working weekends took a lot out of her. I did my usual morning thing: read the paper, played with Nikita, generally putzed around. I then got things ready for my meeting with Mona. I had decided to ride my bike because I live near the Valley Corridor Bikeway, which is a series of interconnected bicycle paths that crisscross a tri-county area of which my town is the hub. The park where Mona had chosen for us to meet sat alongside one of the bike paths. From my apartment I can access the path at one of its rest stops a few miles away. Plus, it was another beautiful day in the neighborhood: sunny, low 70's, light breeze. Coincidence, I'm sure. Yeah, right.

CHAPTER 5

AIR

When I arrived at the park entrance, I didn't see Mona. The only person I saw was a man in his 20's playing Frisbee with his German Shepherd pup. I parked my bike at one of the bike stands and locked it up. I unhooked my pack from the carrier on the back of my bike and slung it over my shoulder. I started to walk toward one of the benches that were scattered around the park when Mona suddenly appeared. She was dressed much the same as the day before, but today she had taken on the Asian look. She smiled at the startled look on my face and said, "Good morning, Michael."

"Good morning, Mona."

"How was your evening with Lily?"

"It was fine, Mona. Thanks for asking." I didn't want to go there so I promptly changed the subject. "How do you do that?"

"Do what? Manage to ask you questions that make you uncomfortable?"

"That too. But, I was referring to your ability to change appearances the way you do, and, for lack of a better term, 'materialize' or 'dematerialize' as the case may be."

"Oh, that," she said with a smile.

"Yes, that."

"Tunneling," she said.

"Tunneling?" I asked.

"Tunneling."

"Okay, I give. What's tunneling?"

"You are fairly well read in the area of physics and quantum mechanics. Surely you have come across tunneling in your studies."

"Come to think of it, I have."

"Did you understand what you read?" she asked.

"I believe so," I said. "My take on it is that at the quantum level a particle can travel from inside an enclosed space to the outside without breaching the barrier in any way. One instant it's on one side of the barrier, the next instant it's on the other side. And, it does this without going over, under, or through the barrier in the conventional sense. The last time I read anything about it, physicists believed the answer was to be found in probability waves, but it was still basically 'unexplained phenomena'."

"You have grasped it quite well."

"So, what does that have to do with you changing appearances or just materializing like you do? Tunneling only occurs at the quantum level, the level of atoms and molecules and quarks."

"Really?"

"You mean to tell me that you can do that? Are you serious?" The look on her face told me that she was quite serious. "How?"

"Like you said, it's 'unexplained phenomena'."

"Come on, Mona, don't give me that. How do you do it?"

"Okay, settle down," Mona said, stifling laughter. "You know how when you go out of body you can be anywhere that you think of instantly?"

"I know of it in theory," I answered. "But, I haven't really been able to do it yet."

"Well, it's like that, but when you 'materialize' somewhere as you like to put it, you take on the appearance of whatever form you visualized before you left."

"Cool," I said, awestruck. I mean, think of the possibilities. "Hey, is it like the intuiting thing?"

"How do you mean?" she said.

"I mean, as I go through this process with you, will I be able to do that, too?"

"Anything is possible, Michael." I didn't know if she was yanking my chain or not. The look on her face gave nothing away.

"Don't go all political-speak on me, Mona."

"You will have the ability to, yes. In reality, you already have it. It's more about knowing that you can, believing it. But, we digress."

"I don't have a problem with digressing. Let's continue to digress for a minute."

"It's infinitely more important to stay on track. Come; let's walk," she said. And, with that she set off through the park making a beeline for the Power Plant that set about a half mile away. I took off after her at a trot. When I caught up, she pointed at the smoke that could be seen pouring out of the stacks at the plant and began her spiel. "Remember when I said that air from the cities could be carried across great distances?"

"Yes," I answered.

"The scientific term used to describe this process is global distillation. When pesticides or other toxic air pollutants are released into the atmosphere in warmer climates, they evaporate and are carried by trade winds to cooler areas where they condense and fall back to earth infecting the air, water, and soil. Not to mention plants, animals, and humans. Your scientists have found traces of insecticides that are used exclusively in tropical areas in trees in the Arctic region."

"Oh, come on." She gave me her 'just how naive are you?' look. "I mean wouldn't it would take a whole lot of chemicals going into the air for it to get from the tropics to the Arctic Circle?"

"Well," Mona responded. "You humans put a whole lot of chemicals into the air. To the tune of over one million pounds per hour." I was in shock at the amount she was talking about.

"And these are pesticides we are talking about?" I asked.

"Not just pesticides. There is also methane from swamps, cattle emissions, and the burning of wood. And, your CFC's (chlorofluorocarbons) from refrigerators, air conditioners, aerosol sprays, and Styrofoam products. Then we have the various toxins that come from burning fossil fuels such as natural gas, coal, and wood, including the forests we talked about earlier. Not to mention the carbon dioxide, nitrous oxide, and sulfur dioxide that your power companies and industrial concerns add to the mix everyday," she said as she glanced toward the Power Plant that was belching smoke. "The top 100 power companies in the U.S. alone account for 90% of the pollution from those three chemicals."

"It takes a lot of energy to run the country."

"But, you don't have to destroy the environment to get it."

"Then what would we use?"

"There are a lot of alternatives that we will talk about later. Again, I want to stay on topic here."

"Okie dokie."

She stopped a few hundred yards from the fence that is meant to keep people from trespassing on the grounds of the Power Company and sat down in the grass. I followed suit. High, wispy clouds began to dot the sky. A lovely counter point to the ugly things we were discussing.

"Air pollution has doubled every ten years since the mid-1800's," she went on. "Pesticide use is up 3,000% since your World War II despite the fact that crop damage from pests is up 27%. The pests are becoming immune to the poisons. Too bad for you that humans aren't. According to your Environmental Defense Fund, Americans breath air 100 times more toxic than the goals set by your Congress ten years ago. This has resulted in a vast majority of your citizens walking around with detectable levels of, among other things, the insecticide chlorpyrifos—a common ingredient in lawn and garden pest controls, indoor foggers, flea collars, and roach and wasp poisons-in their systems."

"You mentioned that air pollution has doubled every ten years since the mid-1800's. Didn't you mention the other day that air pollution had been a problem as far back as the 13th century?"

"In reality, it goes back even further than that. The ancient cities of Greece and Rome had air pollution problems due to the smelting of metals for weapons, coins, and jewelry. But, when we were discussing history, I also stated that the first major problems began during the Industrial Revolution of the 1800's, initially in Europe, then in the U.S., due to the burning of coal and oil. The first modern legislation enacted to deal with air pollution was the Alkali Act of 1863 in the United Kingdom. The first attempts at legislation in the U.S. were municipal regulations in the 1880's. Notice I said municipal regulations. Air pollution at that time was only considered to be a local issue. This is partly explainable due to the lack of technology available at that time. Because of the vastness of the atmosphere the problems of air pollution took decades to manifest and be noticed. But, that excuse won't fly anymore. You now have the technology and you need to start making policy decisions based on long-range predictions."

"Is it really that simple?"

"It's not simple, per se, but doable, yes. You know where the chemicals are coming from and you can base trends on the various types of sources and how they will effect the atmosphere."

"What do you mean by the 'types of sources'?" I asked.

"There are three types of sources classified by your scientists. One is known as a point source. That consists of some of the polluters mentioned previously, specifically the power companies and industrial complexes. It also includes motorized vehicles and concentrated areas of agricultural insecticides and natural sources such as volcanic eruptions and forest fires. The most ominous point source would be a nuclear explosion. Then you have line sources that are simply a number of point sources strung together such as a heavily trafficked highway or a concentration of industrial facilities. Lastly, there is what's known as regional sources which is nothing more than a combination of point and line sources that pollute a large area."

"Unreal," I mumbled.

"Yeah. These different sources are largely responsible for the chemicals spread by the process of global distillation that we mentioned earlier. A large portion of these pollutants comes from 'POP's'."

"POP's?"

"POP's. Persistent organic pollutants. They are called this because they are slow to degrade. And they are very mobile. They are spread around the world by air currents through global distillation and by water currents and through the food chain. These chemicals are insidious because of their semi-volatile nature. This means that they have the ability to present as solids or vapor. They can evaporate in warm conditions, then settle in cool spots where they remain dormant until the temperature rises again which allows them to wake up and resume their global hop-scotching. These chemicals have reeked havoc on life as diverse and remote as fish in the Canadian Arctic lakes, albatross on Midway Island in the Pacific Ocean, and penguins in the Antarctic."

"Pretty much pillar to post, then," I said. I noticed that more people had arrived at the park. An elderly couple was taking a brisk walk along the bike path. Three young women glided past on bicycles. A man and woman were preparing to fly kites with a boy and girl, both toddlers.

"This exposure," Mona continued. "Has led to a large number of human deaths and injuries as well. These chemicals have been implicated in causing liver damage, nervous system disorders, testicular cancer, and reduced sperm count in males, and breast cancer in females. Future generations of humans are being threatened because infants born today have been exposed to chemicals while still in the womb and through breast milk after birth. In mothers classified as meat eaters, testing has shown that 99% of the women had significant levels of DDT in their breast milk. In addition, POP's, along with iodine, pesticides, and lead are implicated in learning and behavioral disorders in one of six children in your country. Traces of Dursban, a chemical that causes nerve damage, have been found in the urine of up to 90% of the children in the U.S. You will hear more about these issues when we cover land degradation." She looked sadly at the couple flying kites with their children before continuing, "Then you have the added calamity of when the POP's finally do begin to degrade, the chemicals that they break down into are worse than the original chemicals and they mix with other POP's to form an even deadlier combo."

"Wonderful. A real witches brew."

"What have you got against witches?"

"Uh, nothing," I said. "It's just an expression. No offense intended to any witches past or present."

"Lousy expression if you ask me." Her talking about the effects of chemicals on little humans had obviously upset her.

"Hey, I won't use it anymore," I said in an effort to placate her.

"I would hope not."

"Anyway," I said wishing to change the subject. "Is all of this we have been talking about with respect to air pollution what has caused the other problems we hear so much about like the hole in the ozone layer, the greenhouse effect, and global warming?"

"In a manner of speaking, but it isn't quite that simple."

"Of course not." *Was anything?*

"As with everything else in the universe," she explained. "It is all connected. Let me start with the so-called 'hole' in the ozone layer."

"Cool."

"Was that a pun?"

"Not intentionally, but now that you mention it."

"Not bad. Anyway, the ozone layer is a gaseous compound made up of three oxygen molecules—O_3. It was formed billions of years ago. One of Spats's (the Spaceman) concoctions."

"Spats (the Spaceman) invented the ozone layer?"

"You doubt my word?"

"No, no, no. Carry on." I didn't want to upset her further. No sense testing her limits.

"I think it would be better if I just let him explain it to you."

"Spats (the Spaceman) is here?" I asked, looking around, my paranoia flaring up once more.

"He will be momentarily. Nothing to be nervous about."

"Easy for you to say."

She fingered her amulet and closed her eyes. I waited quietly. After a moment she opened her eyes and looked past me.

I turned to see a very strange looking man walking towards us out of the woods that bordered the southern edge of the park. He looked like a cross between Albert Einstein and Jimi Hendrix. He was about 5'9" tall and of medium build. At least it looked that way. It was hard to tell because he had on baggy jeans with huge bell-bottoms and an equally baggy sweater over what at one time was a plain white (now gray) dress shirt. On his feet was a pair of worn sandals, no socks. His hair was wild like Albert's, but dark brown like Jimi's and emphasized the ashen pallor of his face. At least what you I could see of his face. He also wore a huge pair of sunglasses that would make Elton John proud. To top it off, he was waving his hands and talking to himself. At least, I assumed he was talking to himself. With all of the intuiting and channeling going on I wasn't sure. Whomever he was talking to, it was very disconcerting. Mona wasn't giving anything away. She was calmly watching him approach. So this was Spats (the Spaceman). Mona had mentioned earlier that he was 'a bit unconventional'. That turned out to be a gross understatement.

He ambled up to Mona who had risen to her feet by this time. They hugged, him still talking non-stop, but now to Mona. I couldn't make out what he was saying. They parted and looked at each other.

"Thanks, you too," Mona said.

Spats (the Spaceman) suddenly went all geek-talking-to-prom-queen shy and blushed a deep red. He managed to mumble, "Ah gosh, Svetladonia."

"Svetladonia?" I crowed as I got to my feet.

"That's my name in a different universe." She turned to Spats (the Spaceman) and said, "We're on Earth in the Milky Way galaxy again, Spats. I'm known as Mona here, remember?"

He blushed even more deeply, which made him look like a red Chia pet. "Uh, sorry Svet...er, Mona."

"That's okay, dear. By the way, this is Michael. Michael, Spats."

I offered my hand. He looked at it, looked at Mona, looked back at my hand. Obviously he wasn't familiar with this Earth custom. I smiled and put my hand in my pocket. He stuck his hand out, smiled, and put his hand in his pocket. He turned his attention back to Mona. That he had a crush on her was quite obvious. I could understand that.

Mona smiled at him and said, "Michael is writing a book on environmental topics with some input from me. I asked you here to help me explain some things to him about the atmosphere and air pollution. Can you do that?"

"Uh, sure, Mona. I'll help anyway I can. Where do you want me to begin?"

"How about with the ozone layer?" Mona said as she sat back down, motioning for me to do the same.

"Uh, okay," he said. He was so intimidated by her I felt sorry for him. But, then he started talking about his area of expertise. The change from a tongue-tied dweeb to a confident lecturer was remarkable. He put his hands behind his back and assumed an erect posture. When he spoke his voice was clear and assured. "The ozone layer," he began. "The function of the ozone layer is to shield the earth's surface and everything on it from excessive, and therefore harmful, ultraviolet (UV) solar radiation. The 'hole' in the ozone layer is a misnomer. In reality it is a thinning of this layer of the stratosphere."

"How does this thinning take place?" I asked.

"Good question. I'm glad you asked that, Michael," Mona said.

"Gee, thanks Kathie Lee," I said.

"Kathie Lee?" asked Spats (the Spaceman). "I thought you were Mona here?"

Mona shot me a look and told Spats (the Spaceman) to ignore me unless I was asking him an intelligent question—something that I would manage to do on occasion. I stuck my tongue out at her and she returned the gesture. We then turned our attention back to Spats (the Spaceman). He stared at us blankly.

"Go on, dear," Mona urged him.

"Uh, okay," he said as he started pacing. "The ozone layer is thinned by a chemical reaction known as a catalytic cycle. The main culprit is chloroflorocarbons—CFC's. That particular compound breaks down into chlorine, florine, and carbon. What happens is the CFC's break up when subjected to UV rays which causes the chlorine to freeze and release the florine and carbon into the atmosphere which in turn causes a chemical reaction with the ozone molecules. Since a radiative balance between ozone and oxygen largely determines the Earth's surface temperature, this results in a fluctuation of said surface temperature. At present, CFC's are responsible for over 50% of the ozone depletion."

"Is that what causes the greenhouse effect?" I asked.

"Partially. As I said, the CFC's account for over 50% of the ozone depletion. The rest of the damage comes from carbon released into the atmosphere when you burn fossil fuels in the form of coal, oil, and trees. And, from methane produced by swamps and rice paddies and, uh, what are those big, lumbering animals that chew their regurgitated matter?" he asked looking at Mona.

"Those would be cows, dear," Mona responded.

"Yes, cows," he said and looked at me for the first time since our introduction. "And, you eat those creations?"

"Not me," I said. "Not anymore anyway."

"I would certainly hope not. How disgusting."

"Let's stay on topic, dear," Mona said.

"Sorry," Spats (the Spaceman) said. "The point I was making is that the majority of the chemicals are introduced into the atmosphere as a result of human consumption."

"The common thread," I mumbled.

"Pardon me?" Spats asked.

"Never mind," said Mona. "Continue please, dear."

"The ozone layer was designed to absorb some of the sun's ultraviolet radiation as it journeyed to the surface of the planet, to let some through to be absorbed by the Earth's surface, with the excess being reflected back into space as infrared radiation re: heat. This balance of absorption and reflection maintains a consistent temperature around the planet. At least that is how I designed it." *He seems a bit miffed. I hope he isn't the quiet, Ted Kaczinski type.*

"Unfortunately," he continued. "With extra chemicals in the form of carbon dioxide, methane, CFC's, and some nitrous oxide being belched into the atmosphere you have created an imbalance. Those chemicals, like ozone, are partially transparent to sunlight. But, unlike ozone, they are not transparent to infrared radiation. So, instead of the infrared radiation reflecting back into space, these other chemicals absorb part of it while another portion is again reflected back to the surface. In other words, the heat is trapped, like in a greenhouse. Hence, the term 'greenhouse effect'."

"Catchy," I said.

Spats (the Spaceman) looked to Mona. She shook her head, signaling him to ignore me again. He shrugged 'okay'. "The problem is getting worse day by day and has been for a long time. Carbon dioxide in the atmosphere is up 30% since the early 1800's and accounts for 50% of the greenhouse gases. It now reaches from the Antarctic region all the way to Chile, Argentina, South Africa, Australia, and New Zealand. At present, you are pumping 5-6 billion tons of CO2 into the atmosphere on a yearly basis. That is approximately one ton for every person on the planet if distributed evenly."

"What do you mean 'if distributed evenly'?" I asked.

"Well, 25% of your world's population—specifically, that of the United States, Canada, Japan, Australia, and Western Europe—are responsible for putting 75% of the CO2 into the atmosphere."

"Oh."

"Yeah, 'oh'," Mona said. "And, tragically, the negotiations aimed at finalizing the treaty to curb global warming that had its inception at the Kyoto Protocol broke down last fall (2000) as a result of last-minute disputes between European and U.S. negotiators. The main bone of contention was the United State's desire to fudge on its responsibility to lower carbon dioxide emissions, a desire fueled—pun fully intended—by financial reasons."

"And, I read the other day that President Bush reneged on his campaign promise to address the issue of carbon dioxide emissions because it would be bad for the economy," I said.

"Yes, I'm sure that came as a huge surprise with his environmental record," Mona said.

"Well, maybe he will change his mind when the economy gets back on its feet," I said.

"Stranger things have happened, I suppose." *No kidding.* "Why don't you continue, dear?" Mona said to Spats (the Spaceman).

"Uh, okay," he said. "The effects of the thinning of the ozone layer and the increase in the greenhouse effect has given rise to the last of the Big Three of Air Pollution: Global Warming. This occurs, obviously, because of the heat being trapped below the stratosphere close to the Earth's surface. Over the past 100 years the mean surface temperature has risen 1.1 degrees Fahrenheit and the sea level has risen eight inches. Since 1990, both Arctic and Antarctic ice have decreased. In the past eight years alone, 7.5 cubic miles of the Antarctic ice sheet has eroded—a warning flag that long-term changes are occurring. El Nino conditions have gotten more severe and have occurred more often than the norm. Also, the oceans have warmed by 0.5 degrees Celsius over the past 50 years. Since 1980 you have suffered through the nine warmest years of the century, five of those in the 1990's. And, as usual you have ignored all of these warnings."

"But, these warnings are all relatively recent," I said.

"Really. Global warming was predicted in 1895 by a human named Suarte Arhelius based on his studies of carbon dioxide emissions. The theory of global warming is that as the CO2 builds up in the atmosphere, which it does yearly, the protons trapped by the greenhouse gases contributes to the increase in ocean temperature which translates to stronger El Nino conditions. As El Nino strengthens it creates massive water pumps that extract water from the earth's surface, transports it, and then deposits it as rain and snow. This, in turn, leads to an increase in floods and blizzards which leads to heat dissipation—a flow to cooler climates—so water vapor from the tropics gets pulled toward the poles which results in more and stronger typhoons and hurricanes and record heat and ice storms. In other parts of the world, the increase in heat results in record droughts bringing about crop failures and famine. So, now you have a positive feedback loop. Warmer climate equals warmer oceans that evaporate more water into the atmosphere since the warmer air has a greater capacity to hold water vapor, which causes warmer oceans, which adds more water to the air and on, and on."

"The heat bone's connected to the water bone; the water bone's connected to the vapor bone," I sang.

"Can't you make him stop?" Spats (the Spaceman) asked Mona. "You have powers."

"It's his attempt at humor," Mona said. "It's a defense mechanism against fear. A human thing, I'm afraid. Just, uh, humor him, okay?" Mona and I both groaned at her pun.

"Oh, my, he's affecting you. You're beginning to frighten me, Mona," Spats (the Spaceman) said.

"Sorry, dear, I'll be more vigilant," she said, again suppressing her laughter.

"Please do. Anyway, this change in climate and its resulting calamities of rain, drought, etc. has affected your food growing systems, too. Cash crops such as tomatoes, soy beans, corn, and wheat have dropped in yield as much as 37% per year over the past decade resulting in lost revenue to the tune of 2-3 billion dollars per year.

"In addition, a myriad of human health issues has arisen because of the Big Three. With only a 1% loss of the ozone layer, the cancer rate has risen as much as 6%. Five hundred thousand of your countrymen develop skin cancer each year. Herpes is on the rise, as are parasitic infections, eye damage, immune system damage, malaria, and cholera.

"Your scientists have said that if the computer models are correct and the temperature trend of the last 20 years continues for another 20 years, it will be warmer than it has been for the past 10,000 years. In fact, the most recent long-term projections call for as much as a 10.5° increase in temperatures over the next century. That could spell catastrophe for life as you know it."

"Your scientists recently discovered," Mona interjected. "That the warmer climate is melting the equivalent of more than 1.25 trillion gallons of water a year from the Greenland Ice Sheet, adding .005 inches to the sea level and increasing the risk of coastal flooding."

"I have asked this many times already," I said. "But, I must ask it again. Why don't we humans do something about it?"

"I don't know anything about that," Spats (the Spaceman) said.

"I do," Mona said. "It's not that you can't. It's that you refuse to. Partly because of political considerations but, mostly because of money. I mean look at the smog in LA or Atlanta. Its man-made, the result of automobiles and industry. Everyone knows it. But is anything being done? Not really. And, why not? Money."

"There has to be more to it than that," I said.

"Does there?" replied Mona. "Well, then." She stopped and got that far away look on her face that could only mean one thing—Lean Gene was lurking. Right on cue her voice changed to that Darth Vader tone, "Hi, Spats."

"Hi Lean Gene," Spats replied. "How come you're channeling instead of being here in person?"

"I don't care to wander among the bewildered herd."

"Ah, I understand," Spats said.

"I knew that you would. Now, I must get on with it. I feel Mona becoming impatient with me."

"Then, by all means, don't let me distract you," Spats said.

Mona/Lean Gene turned back to me and said, "I recently saw an item on your Internet that illustrates my wife's point quite well. The Chairman of one of your major auto makers who shall remain nameless—let's just call him Gourd since that is how hollow his rationalizations are—anyway, this nameless modern human (he said this the way I would say festering boil) freely admitted that one of his products, specifically the SUVs his company manufactures, is bad for the environment. But, he feels obligated to continue producing them because of consumer demand."

"That's Capitalism," I said.

"That's ignorant and irresponsible and downright criminal if you ask me," Lean Gene retorted. "And, on so many levels. From the people who make the product to the people who demand it. A product that even the manufacturer admits guzzles fuel and pollutes the air. Have you all gone mad? Your own scientists know that global warming is mainly a result of industrial pollution and not caused by the sun or other natural causes."

I felt that I should defend my species or, at the very least, my culture. But, I knew what he said was true. I had seen the report that he was referring to in my local newspaper. Nevertheless I shot back, "Bewildered herd?"

"Would you prefer I call you lemmings? Ostriches?" His calmness was irritating.

"I would prefer for you to not be so offensive."

"Yeah, well, I would prefer that you not engage in so much destruction. But, hey, wish in one hand, sh..." Mona shook her head and said in her own voice, "This has gone far enough, I think."

"And, you said he was harmless," I said.

"He is harmless. Open discourse is a necessary part of instituting change."

"I agree, but not when it gets so attacking and personal."

"I know Lean Gene can be a bit, shall we say, abrasive at times, but his point was valid and what's happening to this planet is quite disturbing to those who love it."

"I forget, what exactly was his point?" I asked.

"That until you make environmental products economically competitive or superior to environmentally destructive products—

and lifestyles—nothing will get fixed. You will continue on your merry way to your own destruction."

"What a rosy outlook for the future," I mumbled.

"Yes, isn't it?" replied Mona. We all sat and looked around us at the beautiful trees, the sky, the squirrels, the birds, the other people and their pets enjoying the day. It was disheartening to think of destroying all of this beauty and wonder.

"This would be a good place to stop," Mona suggested. "We have pretty much covered everything that I wanted to with regards to air pollution."

I looked at my watch. It read 4:00 PM. *It's spooky how fast time goes by when I'm with her. The lapses when we watch the other beings in the park or gaze at the sky must last longer than I realize.* As usual Mona read my thoughts.

"It does fly by," she said. "Which is why I am on your case to get this book out."

"Okay, I get it."

"Spats," she said. "It has been an extreme pleasure as always having you around. Thank you so much for coming and for your brilliant lessons and insights into the problems of air pollution here on Earth."

"Uh, er, gee, Mona," he stammered. "I didn't do much." I thought he was going to start scuffing his toe in the dirt. Mona stepped close to him and wrapped him in her arms. He meekly returned her hug. They released each other and stepped apart.

"Have a safe trip home, Spats," Mona said. "And give my best to the rest of the crew."

"I will. We, uh, they…uh, everybody misses you. And, Lean Gene, too."

"We'll be back soon. Michael here is going to quickly hammer out this book, and then we can head home."

"I, uh, we all hope so," Spats said.

"Thanks Spats," I said. "You've really helped me to under-stand air pollution. You are a great teacher."

He mumbled his thanks. He gave Mona one last, long look then started back toward the trees from whence he came. He glanced over his shoulder and threw us a wave. Then he was…gone.

"Boy, does he have it bad for you," I said.

"Oh, he is just so sweet. He is like that around all females." I had actually managed to embarrass her.

"Yeah, I bet."

"Whatever, Michael. We should go if you plan to be back before dinner."

"Okay," I agreed, having had my fun. Plus, I was famished. Some where along the line I had eaten my daily supply of energy bars. We made the trek back to the bike path without much talking. I was lost in my thoughts, mainly about what I had learned today but, also about what to say to Lily tonight and before I knew it, we were back at my bike. I strapped my pack on the rack.

"I know you are very busy during the week," Mona said. "And you will need time to type up what we have gone over so far. So, how about we get together next Saturday morning?"

"Works for me," I replied. "Lily has another program next weekend so she will be tied up, again."

"Yes. God forbid you just tell her the truth."

"Where would you like to meet?" I asked, ignoring her attempt to engage me in a conversation about Lily. "And, what will we be discussing?"

"I would like to meet at Riverside Park downtown," she answered, taking the hint. "We will be discussing water pollu-tion. Say around 10:00 am?"

"Great. I can ride my bike again if the weather cooperates. Think you can take care of that for me?" I said, smiling.

"I'll see what I can do, but you humans have taken a lot of that power out of my hands."

"Well, any help will be appreciated," I said, meaning it on a couple of levels.

"No problem. Have a great week, Michael." She began walking towards the shelter where the restrooms were located and disappeared around the back of the building. I rode home thinking more about what to tell (or not to tell, that is the question...sorry) Lily. I couldn't escape the irony that I was writing a book about, among other things, how to heal the environment, something that should be filling me with a sense of pride and accomplishment, but instead was filling me with guilt because of my refusal to be honest with Lily. I knew deep down that this was most likely going to backfire on me in a big way, but I couldn't bring myself to do the right thing.

INTERLUDE

AVOIDANCE

Lily and I got together for dinner again Sunday night. Since I had told her that I wasn't going to be doing any research I didn't have to lie to her about why I had in fact been with Mona researching air pollution. Instead I had to lie about spending the day hiking alone. The dinner was uneventful, we just went through the motions, mainly due to the fact that she sensed that I was being distant despite my best efforts to act naturally (she can do a bit of intuiting herself). I went home after dropping her at her house. Luckily, that was not out of the ordinary because we rarely spent the night together Sunday through Thursday because of my having to be up at 5:00 am for work and the drive from her house to my apartment took half an hour. Way too long a drive at that time of the morning.

I used my time at work wisely during the week. I did research for Lily on Sustainable Development in the mornings and typed up my conversations with Mona in the afternoons. Lily and I talked on the telephone quite frequently per usual and she seemed pleased with the amount of research I was getting done on her behalf. But, she could sense that something was up and asked me a few times if something was bothering me. I maintained my denial, stubbornly refusing to come clean. I felt like one of those guys the mobsters make dig their own grave before burying them in it.

Friday rolled around and, mercifully, Lily had a bachelorette party to go to for one of her girlfriends. This meant I didn't have to lie to her face-to-face for another day. On the phone is hard enough, but in person is exhausting. I would never make it as a politician…maybe a lawyer, but definitely not a politician. I had survived so far by reminding myself that Lily had lied to me once. (It concerned a not-so-old boyfriend, a bouquet of flowers, and a less-than-convincing alibi as to her whereabouts one Saturday night—but that's all I'm saying about that.) I know that two wrongs don't make a right. But, they do provide a nice basis for rationalization. And that's all I had at the moment.

We spoke briefly before she went out for the evening and made plans to see each other Saturday night provided she didn't run off with the male stripper from the party (Ha ha—her joke, not mine). I used my free Friday night for some pleasure reading and to watch a little TV. And, to play with Nikita the Cat, of course.

CHAPTER 6

WATER

I was up early as usual Saturday morning—thanks to the afore-mentioned cat—and was able to get in a little extra meditation time before I left for my meeting with Mona. She had come through with another beautiful spring day and I opted to ride my bike again. I took one of the connector paths of the Valley Corridor bike trail that runs parallel to the river and made it to our meeting location—Riverside Park in the downtown area—about 15 minutes early.

Mona was already there, sitting on a rock at the water's edge mediating. She had reverted back to her original blonde/blue-eyed look. I wondered if she had been there all night or was just habitually early. I stopped my bike about 20 yards away from her and waited for her to acknowledge my presence. After a few minutes, she stretched like a cat after a nap and gracefully slid off the rock. Turning to me, she said, "Good morning, Michael."

"Morning, Mona," I replied. "Thanks for the beautiful weather."

"Not entirely my doing, but you're welcome anyway," she replied. "Riding your bike, again I see. How wonderful. I dare say you are taking our talks to heart."

"I hate to burst your bubble, but I have been riding a bike for years. And, much like today, it's more to keep my svelte boyish figure, rather than save the planet." I grabbed my pack off the bike and sat it next to a big rock close to the one Mona had been using.

"Just so you do the right thing, it doesn't matter why you are doing it. If everyone waited for the right motives to do anything, the world would end up…okay, bad example. But, you get where I'm going."

"Beautiful spot." *Let's not get into the heavy stuff yet.*

"Yes, it is," Mona said. She glanced around. "Or, it would be anyway. You know, if it weren't for the plastic pop bottles floating in the river and the aluminum cans along the bank." *So much for not getting into the heavy stuff yet.*

"Yeah, there is that," I agreed.

"Do you have any idea how important water is to the survival of virtually all life on this planet?"

"Well, I know that the human body is made up of a large percentage of water. I'm not exactly sure how much, but I know it's a lot."

"The human body is 65% water," she said. "The blood that flows through your veins is 90% water. If you humans lose just 12% of your body water at one time you risk death. You need 2 to 3 quarts of water per day to maintain that svelte boyish figure of yours. With six billion people on the planet at present, that is between 3 million and 4.5 million gallons needed per day."

"That's a lot of water."

"Yes, it is. And, that's just drinking water for humans. Now factor in showers and baths, brushing your teeth, shaving, doing dishes and laundry, watering the lawn, washing the car, house cleaning, cooking, and what's needed for industry and animals—the list could go on and on. At present rate, 20 billion gallons more water is pulled from the ground each day than is restored by rainfall."

"Twenty billion gallons?"

"Twenty billion gallons."

"Per day?"

"Per day."

I leaned back against my rock and shook my head in amazement. She smiled and sat down on her rock. I got out my tape recorder and switched it on.

"This is an extremely important statistic," she said. "Of the approximately 326 million cubic miles of water on Earth, 97% is salty."

"Jeezy Pete."

"Jeezy Pete? You writers do have a way with words."

"Thank you. A moment ago you said 'out of the ground'. What about lakes and rivers and such?"

"Groundwater alone accounts for 50% of your drinking water," she said.

"When you say groundwater, exactly what do you mean?"

"I'm going to let Dr. Don answer that."

"Fine, but when is going to get here?" I asked.

She smiled and nodded as she looked past my shoulder. I turned around to find a very tall, lean man standing a few feet behind me. He was dressed elegantly in a white suit of a cloth I had never seen before. The suit jacket covered an aquamarine tee shirt of the same material. He wore white cloth loafers, sans socks. His angular, clean-shaven face was framed by a mop of hair so bleached it was almost white. His eyes matched the color of his tee shirt and sparkled with intelligence, humor, and a hint of arrogance. I suppose most women would describe him as to-die-for handsome. I stood up.

"Michael, this is Dr. Donald Aquilerre. Dr. Don, Michael," Mona said.

We shook hands. He showed none of the social ineptitude of Spats (the Spaceman). In fact, he was the embodiment of suave.

"Wonderful to meet you, Michael", he said. "I've heard so much about you." His voice, smooth and rich, put me in the mind of expensive milk chocolate.

"Really. Like what?" I asked.

"Oh, not to worry. It was all very complimentary. Well, except for Lean Gene's comments, but he doesn't like any of you humans as a rule, so I wouldn't put too much stock in that."

"Just what did Lean Gene have to say about me?"

He made a brushing aside motion with his hand. "Nothing worth repeating." *Yeah, I bet.* He turned to Mona. She offered her hand as she stepped forward. He took it and brought it to his lips for a brief kiss, more of a dusting of the back of her hand. "Your most alluring disguise yet, Mona," he said. "But, still no match for your true form."

"Thank you, Donald. It was very kind of you to come."

"I'm always at your beck and call. You know that, Mona. So, I take it you require my expertise in explaining the hydrological workings of this quaint little satellite of that gaseous monstrosity known as 'sol'?"

"Yes. Spats (the Spaceman) enlightened us on the intricacies of the atmosphere yesterday and I thought it would be nice if you would do the same with the waters of this world. After all, you know the subject so much better than I do."

"Mona, you are too kind. But, I feel that you exaggerate. You know this stuff as well as I do, I'm sure."

"Now it is you who are too kind. Put away your veil of modesty and you will have to admit that no one knows this subject as well as you."

"Mona, I again must once again protest."

"Oh, for God's sakes. Could we get on with it?" I interrupted. The love fest was making me nauseous.

They both looked at me with shock for a moment before Mona said, "Forgive him, Donald. These humans—the ones of this culture anyway—are always in such a hurry, usually to go nowhere. They don't see that that is a big part of their problem."

"It's quite all right, Mona," Dr. Don replied. Mona looked at me like I was gum on her shoe.

"I'm sorry," I said. "That was inappropriate."

"No need to apologize," Dr. Don said.

"Easy for you to say." I nodded towards Mona.

He smiled and nodded knowingly. "Perhaps we should, er, 'get on with it' as you say."

"Wonderful idea, Donald," Mona said, sporting a warm smile that turned to ice when she swung head my way.

"Where would you like me to begin?"

"With an explanation of groundwater, if you don't mind," Mona said.

"Very well. Groundwater it is then." He gestured for us to sit down which we did. He then began to pace while he talked. "Groundwater is water that occurs in saturated, non-consolidated geologic material." He paused when he noticed the blank look on my face. He looked to Mona.

"He's a novice," she said.

"He knows nothing?" Don asked.

"It's easier to assume that, yes."

"Excuse me,' I said. "But I'm sitting here." They both smiled, a little too pityingly for my taste.

"In laymen's terms," Dr. Don continued. "The geologic material that I referred to is sand, gravel, and porous rock. These saturated strata are called aquifers. Fully 97% of the Earth's groundwater is stored in aquifers. And, groundwater makes up almost half the water in the hydrological cycle." He paused, assuming my ignorance of the topic.

"What is the hydrological cycle?" I asked, hating to disappoint him. Plus, I was ignorant on the topic. Damn it.

"The hydrological cycle is the movement of water between the atmosphere and the land, the oceans, lakes, rivers, and ponds. By movement I mean the process of rainfall and snowfall into these bodies of water and the resulting evaporation and condensation which leads to more rainfall and snowfall. Without water there would not be life on this planet. Hence, without water life as you know it will not survive."

"Okay, I get how important water in general and groundwater in particular is to life and where over-population is putting a strain on the supply, but where does pollution fit into all of this?" I asked.

"Where to start?" He paced some more. "I guess we should start in the middle and work our way out. The pollution of the various water systems comes from a wide variety of sources. You have already heard about acid rain in your lecture on air pollution." He stopped when he saw me frantically waving my hand and arched his eyebrows.

"Acid rain?" I asked. He looked at Mona.

"Spats (the Spaceman) didn't cover acid rain in his lecture?" he asked.

"I'm afraid not," she replied. He turned back to me.

"And, you don't know what acid rain is?"

"I'm afraid not," I said, mimicking Mona which earned me yet another of her icy stares. I ignored her…for the most part.

"Very well," said Don. "Then I believe a brief tutorial on acid rain is in order. Natural rain has a pH of around 5.7 acid content." My hand was in his face again. He held up his hands to halt me before I could get the question out. "The measure of a substance's acidity is known as it's pH. In turn, pH stands for 'potential of hydrogen' and is the hydrogen ion—a positively charged hydrogen atom—concentration of a solution. The lower the pH, the more acidity. It is measured on a logarithmic scale and goes in descending order. Each change in whole number represents a factor of ten. Thus, a pH of 5 is ten times more acidic than a pH of 6 which is ten times more acidic than a pH of 7 and so on." He paused to see if I was absorbing this.

"I'm not the village idiot, okay," I said, tiring of wearing the dunce cap. "It's just that I have never studied this particular subject matter before. I think you will find I am a quick learner and you won't have to repeat yourself."

"My apologies."

I nodded my acceptance.

"As I stated previously," he continued. "Natural rain has a pH of approximately 5.7. The worst acid rain usually measures between 3.7 and 4.7. So acid rain is anywhere from 10 to 100 times more acidic than natural rain."

"How does acid rain become acid rain?"

"Humans interfere with nature in an inappropriate manner." I couldn't tell if he was being a smart aleck or just stating the facts. I gave him the benefit of the doubt. As yet, he hadn't demonstrated an inclination towards humor.

"All right. But, how exactly?" I asked.

"By pouring sulfur dioxide into the atmosphere."

"I thought sulfur dioxide occurred in the atmosphere naturally from volcanoes and decaying matter." *I don't remember where I read that, but I am so pleased to have recalled it now.*

He looked at me with a hint of respect and said, "It does. And, when that was the only source, nature was able to deal with it as a part of the hydrological cycle. It would enter the atmosphere, combine with water, get rained back out, and return to the planet's surface. But, when humans began burning fossil fuels in great abundance, particularly coal, all that changed. Humans pump about 100 million tons of sulfur dioxide into the atmosphere yearly. That is a number roughly equal to the natural contribution. So, humans have a 100% impact on the environment in this particular area of destruction.

"The result of this is fish kills in your rivers and lakes. Fully 20% of your freshwater fish have been either harvested or polluted into extinction. Other problems are the seepage of toxic chemicals into the aquifers; nutrient depletion and contamination of the soil which leads to a decline in the growth of plants and trees with the domino effect of massive animal and insect kills; the corrosion of products made of metal or paint from edifices which also seep into the water supply; and, last but not least certainly from your standpoint, human health problems in the form of gastrointestinal disorders, respiratory problems, and heart disease just to name a few."

"Gee, that sounds groovy," I said. He looked to Mona for a translation.

"Like I counseled Spats (the Spaceman)," she said. "It's best to ignore comments like that and only acknowledge legitimate questions that he occasionally manages to come up with."

"Thanks," he said and began to pace again. "LUSTS are another major source of groundwater pollution."

"LUSTS?" I asked.

"LUSTS. Leaking Underground Storage Tanks. The tanks usually hold gasoline. And about three million gallons of it leaks into the ground each year. A lot of it ends up in aquifers. Out of approximately 100,000 commercially owned tanks that have been detected as having leaks, 18,000 or so are known to pollute groundwater systems. In Texas alone almost 90% of counties have reported LUSTS. Silicon Valley in California has reported an 85% leakage rate in its tanks and has more Superfund clean-up sites than any city of its size in the U.S. There is a price to pay for your computers. The problem with these LUSTS is that one gallon of gas per day infiltrating a water supply system can taint the water of a city of 50,000 people.

"Then you have your farmers and ranchers, both commercial and private. More water pollution is attributed to these two sectors in the form of pesticides and animal waste than all municipal and industrial sources combined."

"Oh, come on. How can farmers and ranchers out-pollute industry?" I exclaimed.

"Easy. For starters, farm animal waste is produced at 130 times the rate of human waste and you have a lot of animals out there in order to feed your culture's meat obsession. But, I'll not get into that particular issue here. Then we get into the toxic chemicals. Pesticide, insecticide, and herbicide runoff in particular. Those three sources contribute 1,400 different active ingredients of chemicals into the environment, which usually end up in your streams, lakes, rivers, aquifers, and wells. Over 700 different toxic compounds have been drawn from U.S. water systems alone. I am talking about chemicals such as chlordane, diazonin, dioxin, sodium, sulfate, and the ever-popular DDT."

"Wait a minute," I said. "I thought DDT was banned years ago."

"It was. Thirty years ago, but it still shows up in your water supplies. It has a long half-life as do a lot of the other chemicals I mentioned."

"That sucks," I said.

"Quite," he agreed. "And, in addition to the agriculturally produced pollution, we have the industrially induced pollution to talk about. From industry we get such wonderful substances as benzene, aluminum, arsenic, copper, lead, and mercury. Plus, from nuclear facilities comes radioactive radium and uranium. And, from the water treatment plants themselves asbestos, chlorine, fluoride, and ozone."

"From the water treatment plants themselves?" I asked.

"Yes," he responded. "Sometimes the chemicals that are used to treat the chemicals that need treated either leak themselves or the combination of the two chemicals results in yet another dangerous compound."

"No wonder bottled water has become such big business," I mumbled.

"Another issue," he continued. "Is that a great deal of water is needed just to produce your consumer goods and such. It takes 4,000 gallons of water to produce the steel for one automobile, 50,000 gallons to produce rayon carpeting, and 700 gallons to generate one kilowatt of electricity at a coal fired power plant."

"So it's mainly a problem of the northern industrialized states?" I asked.

"Hardly," he answered.

"Meaning?"

"Speaking strictly of your United States?"

"Yes."

"There are different sources from different parts of the country. The Northwest has industry; the Midwest, primarily agriculture; the Southwest, mining and the defense industry; the West, plastics and computer products; the South, petrochemicals and agriculture; and the East Coast, industry and poultry farming."

"What kinds of damage is all of this pollution causing besides screwing up the drinking water?"

"Get comfortable, we're going to be here a while."

"Wonderful."

"And, note that I can't restrict my comments to just your country because of the nature of the problem."

"So noted, counselor," I said. Mona and I exchanged a smile. Dr. Don ignored us.

"Okay. First of all," he went on. "Your rivers and streams and lakes are being ruined. The Mighty Mississippi is filthy thanks to oil refineries and chemical plants. Certain species of fish, particularly Chinook in the Snake River and White Croaker in various California rivers are facing extinction because their habitats are so polluted. In Europe, the Danube River and the Rhine are open sewers; the Elbe River in Germany is flooded with mercury; in India, the Ganges is in a state of total decay; and in what was Czechoslovakia, 70% of the rivers are heavily polluted.

"As for lakes, your Great Lakes are heavily polluted due to shipping traffic and airborne chemicals blown in from industry to their west and Lake Bailal in Russia, the largest aggregation of fresh water on the planet, is beginning to suffer major fish kills. Your Environmental Protection Agency has found that as many as 20,000 streams and lakes don't meet water quality standards—that constitutes 40% of the streams and lakes nationwide. And, to compound the problem, a program to deal with this issue was recently voted down in your Senate."

"Somehow that doesn't surprise me," I said.

"That's a big part of the problem," Mona said. "Nothing much shocks the people of your culture anymore. And, when something does get your attention, most of you are either too afraid or too apathetic or too self-absorbed to do anything about it anyway."

"Unfortunately I can't deny that," I said. "But, the key word is 'most'. That means that there are a few who do choose to get involved and take action. And, that is a reason for hope."

"Why, Michael," Mona said, smiling. "I believe that you are truly grasping the heart of this message—in spite of all that is wrong with the planet, and many of it's inhabitants, there still is hope."

"Why thank you, Mona," I said. She nodded to acknowledge having intuited how backhanded I considered her compliment to be.

"If you two are done with your, what was the term you thought of, Michael? Oh, yes, your 'love fest' I will continue," Dr. Don said actually displaying a sense of humor. And, an ability to intuit.

"By all means, Donald," Mona said.

"Very good. Another pollution source is dams. You stop the natural flow of the river and harness it's power to produce electricity in hydroelectric plants and call that environmental progress when in fact the combination of decaying vegetation and methane-producing mud on the valley floor brought about by the damming puts just as much greenhouse gas into the atmosphere as if you would have gone ahead and burned coal. Japan alone has built 1,000 dams since 1950 with 500 more on the drawing board. Dams have fragmented fully 60% of the world's largest rivers. The carbon dioxide produced as a result of these dams accounts for the third highest amount in the atmosphere behind power companies and automobiles. And, the polluted rivers that aren't dammed up pollute the estuaries that they flow into."

"What are estuaries?" I asked. *A part of the female anatomy that I haven't heard of?*

"Estuaries are the wide mouths of a river where its current meets the ocean tide. At the confluence are a combination of wetlands, river water, ocean water, and marshes. It is a very delicately balanced system. In combination with the moon's gravitational pull it draws the river's decaying organic matter and various minerals into the ocean to mix with the 3% salt solution. The result being to form a nutrient rich aquatic food web of plankton, algae, and bacteria that in turn attracts insects, flies, crabs, snails, oysters, clams, and small fish. These smaller prey attract ospreys, hawks, and larger fish and on up the food chain to humans. You can just imagine how much havoc a disruption of any link in the chain could cause. In the last one hundred years you have lost almost half of the world's wetlands. Man made pollution in the Chesapeake Bay in Virginia has activated long dormant bacteria that are now causing fish kills and diseases in humans.

"In addition, due to the dumping of sewage from ships, offshore pollution, a huge coastal population, and oil spills—of which there are between 10,000 and 16,000 reported each year—marine life is being threatened in a major way. Mainly because 90% of marine life is coastal and 90% of the pollutants come from land based activities. The beginning in this terrible chain of the destruction of marine life is the fact that plant plankton is being wiped out. The danger comes from the fact that plant plankton is the basic food of marine life. It also performs the vital function of photosynthesis—oxygen production for the, uh, layman." He smiled when he said that last bit. I returned his smile. Mine wasn't as genuine.

"The plant plankton feeds animal plankton," he continued. "Which feeds small fish, the small fish act as food for the larger fish, etc., etc. Along with the destruction of the plant plankton, coral reefs are being decimated and they are a mini-ecosystem in their own right supporting a wide variety of marine life. The problem with a huge coastal population is that it instigates construction of such things as harbor jetties, homes and businesses, seawalls, and breakwaters that all add to pollution and cause coastal erosion. Those calamities combine to produce such disasters as red tides in Florida; 7,000 beach closings in the U.S. in 1999 of which 70% were attributable to bacteria; green slime in the Adriatic Sea; the closing of beaches in Australia two or three days per week; the introduction of toxic chemicals in marine life such as dolphins off the New Jersey coast with 3,000 times the government limit of PCB's in their systems; fish, sea birds, and turtles dying from eating discarded plastic products; and the decline in population of various food fish in California fisheries by 90% since 1982."

"That is tragic," I said.

"Oh, wait; let me get to the specifics of what water pollution does to humans."

"Let's hear it." *May as well get it over with.*

"Your doctors say that water pollution is the primary, I repeat primary, source of disease in the United States. Worldwide, four out of ten people drink and/or bath in contaminated, disease carrying water. Some of the diseases implicated are colon and rectal cancer, kidney disease, osteoporosis, hypertension, blue-baby syndrome, birth defects, dermatitis, and various and sundry viruses. Not to mention the starvation of one billion people because of poor irrigation practices and contaminated water supplies while the Western cultures waste the water that they haven't managed to pollute yet."

"Waste it how?" I asked.

"By your ridiculous dietary habits mainly. Globally you humans use 65% of your water just to feed your masses leaving only 35% to take care of all the other water needs mentioned earlier combined. In California, farmers use 85% of the water. The majority of that is used to irrigate pastureland for cattle. That is more water than is used by the populations of Los Angeles and San Francisco combined. It is estimated that it takes 1½ million gallons of water to produce the 1,500 pounds of food eaten by the average American household per year."

"We will get into that more in our next talk," Mona interrupted. "Which will be about land-based issues, including food production problems."

"Good," Dr. Don said. "That is not my area of expertise anyway. In conclusion, let me say that once you use up your groundwater, you are going to be in dire straits. Especially since you are at the same time continuing to pollute your other water systems and grow your population unchecked. It takes 1,400 years for groundwater to recycle and you don't have that kind of time."

"Not even close," Mona said. "If things keep going the way they are now."

"Do you have any questions, Michael?" Dr. Don asked.

"Not at the moment," I answered. "But, I'm sure I will after I have listened to this tape again."

"Well, if you do," he said, "I am confident that Mona will be able to answer them."

"Thank you for your insights, Donald," Mona said.

"Always a pleasure to see you under any circumstances, my dear. I hope that I was of some help."

"You were," I interjected. "Very much so. Thank you."

"You're welcome. But, time will tell if what I had to say makes any impact. From what I have seen of your culture and those like yours, I don't hold out much hope for the survival of your species. I hope I am wrong."

"I believe you are," I said. "The human spirit and the will and call to action that my species is capable of can be very impressive. If we put our collective minds to it, we will see our way out of this mess."

"I sincerely hope so," he said. "But, that's a mighty big 'if'. At any rate, I bid you a fond farewell."

He reached out for Mona's hand, but she shoved it aside and moved in to give him a fierce hug. She said, "Take care of yourself, Donald. And, thanks again."

They moved apart and he said, "I will, Mona, you, too."

With that, he walked off down the bike path towards the boat ramps although there were no boats docked there. While they were saying their good byes, I had looked at my watch and was, per usual, amazed that it was already late afternoon. Mona tapped me on the shoulder and I turned my attention back to her.

"Pretty interesting fellow, no?" she asked.

"Yes, very interesting." I glanced over my shoulder to have one last look at Dr. Don, but, of course, he was gone. I must have been getting used to the tunneling because I wasn't even surprised.

Mona and I made small talk for a while. Finally she said, "I should be going. I will contact you as to when and where to have our conversation about the land. But, before I go I have something I want you to think about."

"Like I don't have enough already," I said.

She smiled and said, "True, but I'm certain that you can handle one more thing."

"I imagine I can."

"Good. You mentioned to Donald about your species putting your 'collective minds to it', remember?"

"Yes."

"As you know everything, including thoughts, is pure energy. Therefore, it would follow that everything is connected in some-way."

"We've been over this. Your point?"

"My point is that thoughts are one of the building blocks of the universe," she responded.

"I'm not following."

"Okay, how to make it more clear?" She paused and stared off into space. I thought for a moment that Lean Gene was about to show up again. But, I was spared that when she said, "You know those experiments in quantum mechanics where just observing a particle can change its behavior?"

"Yes," I said.

"And, how if you split an electron in two, send the two halves millions of miles apart, and change the direction of one half and the other half responds in the same way?"

"Yes, again," I answered.

"Well, then wouldn't it make sense that if enough of you humans change the direction of your thoughts that you could influence other humans to change the direction of their thoughts in the same way?"

"You are speaking of what I call the 'collective consciousness'?"

"Yes. And, I am proposing using it to change the Gestalt of your culture to one of an environmentally responsible one. Possible you think?"

"I guess so."

"Think about it, okay?" Mona asked.

"I will," I promised.

"Wonderful."

I had gathered up my things while Mona had been espousing her ideas on the 'collective consciousness'. We walked to where my bike was locked up.

"How about tomorrow we meet at the Organic Garden Co-op at the same time we met today?"

"Great, I can ride my bike again. The bike trail has a connector in that area."

"Yes, I know," she said, smiling.

"Of course you do," I said, smiling myself. "Because of the choice of meeting sites for tomorrow, I take it we will be discussing the food production issues that you and Dr. Don mentioned?"

"Among other things," Mona answered. "That will be at the end of the day. First we must cover the destruction of the land that is taking place."

"Oh, boy, another low-stress day," I said. I strapped my pack to my bike and straddled the seat.

"Well, if it makes you feel any better tomorrow is the last day where we focus on the problem. Next weekend we will be focusing on solutions."

"Really? Wow, this is moving along quickly."

"Again, by necessity if it is to do any good. Be careful riding home, Michael."

"Thank you, Mona, I will. See you tomorrow. Tell Lean Gene I said 'howdy'."

"I most certainly will," she said. "That's very sweet of you." Then, she kissed me on the check. I rode off, blushing. I didn't bother to look back, figuring that she would be gone already. But, I was wrong, as I was soon to find out.

INTERLUDE

BUSTED

When I arrived at Lily's to pick her up for dinner, all hell broke loose. Turns out Susan Buckley, a friend of hers, had been downtown earlier in the day and had spotted Mona and I a few minutes before we ended our meeting for the day. Which means that she saw Mona's kiss on the cheek. Susan told Lily that as soon as Mona had seen Susan watching her, she slipped off behind the restroom area in the park, then simply disappeared (I couldn't wait to explain to Lily what *really* happened). Between the kiss on the cheek, Mona's suspicious behavior, and my recent 'emotional distance' Lily put two and two (or, in this case three and three) together and came up with, "You cheating bastard." I figured at that point my only recourse was to tell her the truth. About Mona, me writing a book, the intuiting, the tunneling... all of it. She would believe me, right?

"Mother Nature," she screamed. "That's the best you can come up with? If you've met someone else just tell me, but don't treat me like I'm an idiot. Mother Nature, for God's sakes."

"I know it's pretty far fetched," I said. "But."

"Far fetched! You get caught cheating on me." she started, but I interrupted with, "I'm not cheating on you." which she trumped with, "Don't interrupt me, I wasn't finished yet."

"Sorry." I slumped down on the couch while she stalked the living room like a lioness sizing up her prey.

"You get caught cheating on me," she repeated and pointed at me, glaring, when I started to defend myself again. "And, in defense you have the gall to offer up the explanation that you are not cheating, you are doing research for a book on the environment at the behest of and with the help of a woman—who just happens to be beautiful, I might add—and, who you say is Mother Nature. Do you realize how inadequate 'far fetched' comes to covering this?"

"Yes, I do, but it's true nonetheless," I said. "It was difficult for me to believe at first, too, but Mona convinced me."

"Aha, Mona! I thought she was Mother Nature," she snarled.

"Mona is the name she uses when she." I hesitated trying to decide how to say this.

"When she what?" Lily asked.

"When she is in this solar system," I said, deciding it would be the truth, the whole truth, and nothing but the truth from now on.

Lily stared at me for a moment, then said, "Is it booze? Is that it, you're drinking again?"

"No, I'm not drinking again." *Although about now it doesn't sound like such a bad idea.*

"Well, if you're not drinking, then I guess you think I'm incredibly gullible...or incredibly stupid. Either way, that really ticks me off." She stopped next to her desk and picked up a tape dispenser.

"I don't think you are stupid at all," I said.

"So, gullible it is." I hesitated before answering; after all she was gullible at times. But, I didn't feel that this was the best time to let her know that—I mean, now that she was armed. The hesitation thing was killing me, though. "You son of a," she began.

"If you would only let me tell you the whole story," I pleaded.

"And, I'm sure a story is exactly what it would be, too."

"Lily, if I've been willing so far to tell you that I've met Mother Nature, *the* Mother Nature, and that she's here from another solar system, then doesn't it stand to reason that I might as well come clean with all of it?"

"Okay, Mr. Truthful, Susan already gave me a general rundown on what she looks like." *I bet that Susan gave details that a veteran CIA operative wouldn't be able to match.* "But, she couldn't get a fix on her age. Something about her appearance seeming to change. So, how old is she?" *Here we go again.*

"Twenty billion years." I ducked, and the tape dispenser bounced off of my shoulder. "Ow. Be careful. You could put someone's eye out," I said trying for some levity.

"Be grateful it wasn't the letter opener," she said as she stalked back to her desk—to reload I presumed. *So much for levity.* I got up and moved behind the couch. I wanted some protection. And some maneuverability.

"I can't help it if she says she's twenty billion years old. Besides, she's married."

"That's funny, Susan said she wasn't wearing a wedding ring." I told you. Susan probably could describe the tread on the bottom of Mona's shoes.

"I asked her about that," I tried to explain.

"I just bet you did," Lily interjected.

"She said that where she comes from they don't practice that custom."

"Really? How convenient. And, where is this husband of hers while she is sneaking around with you?" I didn't bother to correct the 'sneaking around' shot.

"She said he stays at the house they are living in while on this planet," I said, trailing off as I got to the end of the sentence. She just shook her head as I continued to dig myself in deeper.

"So, you haven't met him then, I presume?" she said.

"Well, not exactly." I was leery of telling her the rest, especially since she now had a stapler in her hand.

"Not exactly?"

"I've heard his voice," I said. She raised her eyebrows questioningly, awaiting further explanation. "He offers his opinion on various topics now and then through Mona."

"What, she channels him?" she said. She could do sarcastic, too. I eyed the stapler as I nodded my assent. "You're telling me that Mother Nature channels her husband?" I nodded again. "Okay," she said pursing her lips indicating that she was really getting fed up. "And, to whom pray tell would Mother Nature be married?" *I was afraid you were going to ask that.* I raised up on the balls of my feet, ready to dodge either way when she unleashed the stapler. "Father Time," I said. The stapler just missed my head as I ducked behind the couch. *This isn't going well.*

"You are really getting me angry, now, Michael," she said. *Gee, really?*

"You need to take a leap of faith here, Lily," I said, peeking around the end of the couch.

"A leap of faith? Really? I need to take a leap of faith?" She was looking around for something else to throw. She opted for a small, but heavy paperweight. "Okay, here I go." She jumped up and down once. "I took my leap of faith, Michael. And, I landed right back here on the nice solid earth. Care to join me?" She motioned me over. "Come back from whatever planet you're visiting?" I shook my head 'no'. *You're just trying to lure me closer for a better shot.*

"Come on, Lily, settle down for a minute, huh?"

"Settle down!" She was edging closer to where I was hunkered down behind the couch. I kept my eye on the paperweight.

"This is getting ridiculous," I said. *Whoops.*

"You think? Gee, you tell me Mother Nature and her husband Father Time—by the way, does he have a nickname too, or does he go around telling people his name is Father Time? Assuming there is a husband at all, of course."

"His nickname is Lean Gene," I said.

"Lean Gene?" she said.

"It seems he used to be rather heavy at one time." I said, willing to tell her everything, but she held up her hand to stop me.

"I am tired of listening to this, this…fantasy of yours," she said.

"Trust me, Lily." she gave me a look worthy of Mona. "Okay, bad choice of words. I know how crazy all of this sounds. I had trouble believing her when she told me. But, I'm convinced she is who she says she is."

"And just how did she convince you? Never mind, I don't think I want to know the answer to that," she said.

"She asked me to keep an open mind and reminded me of beliefs that I supposedly held about past lives, other dimensions, spirit guides, collective consciousness—all things that you believe in, too, Lily."

"Don't try to twist this around on me, Michael."

"I'm not. The only reason I didn't tell you about this in the first place is that I know how outlandish it all sounds. I just wanted to wait until I was done with the book, or at least the research, before I told you everything. I felt it would be a lot less complicated that way."

"Well, you figured wrong, didn't you?"

"Yes, I did," I admitted. "And, I'm sorry. I admit that I didn't handle this well at all."

"It's always hard admitting to betraying a trust," she said.

I heard Mona's words in what Lily had just said. I had meant well, but had taken the easier, softer way and it had backfired on me. I had lost my credibility with Lily and it was going to be an uphill battle to regain her trust. "I should have told you the truth from the start," I said. "And why I didn't is irrelevant. From now on I promise to tell you everything."

"And I'm supposed to believe you now? It's not that simple, Michael."

"I know, but it's all I have."

She stared at me for a moment, then the anger drained from her face. "I need some time to digest this. You need to go." She put the paperweight back on the desk.

"Lily," I began in an attempt to assuage my feelings of guilt and shame more than anything. But, she held up her hand again to quiet me.

"Just go, Michael," her voice more adamant. So, I went.

Home. I thought about what had taken place over the last few weeks. I came to the conclusion that I had acted out of fear. I had been a coward, rationalizing my lying as the best way to handle the situation, when in fact it had just been the easiest way. The old 'ends justify the means' rationalization. I could relate to the people of both the 'left' and the 'right' who meant well, but somewhere along the line lost their way and resorted to drastic measures to further their agenda. I abhorred that mentality when I witnessed it in others—be it an individual, a corporation, or a government. That kind of thinking was arrogant in that it indicated a lack of trust in the individual or the public to make intelligent choices for themselves. And, I was guilty of the same arrogant behavior.

My mulling over of this problem eventually took my thoughts to the conundrum that Mona had put to me earlier about 'collective consciousness'. I had put dishonesty into the universe. How did I expect others to act courageously, face their fears, work to change the status quo if I couldn't do it myself? I had the revelation that it was individuals thinking and acting like I had that had gotten the planet into the mess it was in now. I more fully understood what Mona was talking about when she said that this type of thinking and acting had led the majority of the human race away from nature and spirituality. And how that opened the door for the mentality of 'ours is the only right way' and 'the earth and everything on it is for us to use as we see fit'. I saw that the trend of fear and selfishness and exclusion and anger that currently ruled cultures such as ours needed to be reversed. I resolved that from now on I would be honest and let the chips fall where they may. I would try my hardest to contribute positive energy to the collective consciousness.

All this had left me exhausted and I finally drifted off to sleep. Nikita woke me per usual around 5:00 am. I went through my morning routine and got it together to meet Mona. I was tempted to call Lily before I left, but decided that was probably not the best of ideas. I strapped my pack to my bike and took off for the garden co-op. Mona had ordered up yet another beautiful spring day. If she stayed around, I wouldn't need to move to Arizona.

CHAPTER 7

LAND

I had decided to tell Mona about what had transpired between Lily and I and get some feedback on how best to handle it. I figured a woman's perspective wouldn't hurt. But, to my surprise, when I arrived she was not sitting and meditating, nor was she alone—and now she was a Native American. She was leaning with her elbows on the rail of the fence that surrounded the garden talking with another woman who, from the side, looked vaguely familiar.

I dismounted my bike and leaned it against the fence. Extracting my lock and chain from my pack, I secured it to one of the fence posts. I couldn't help but notice that Mona glanced at her companion as if to say 'These poor humans'. I ignored her. I had lived in the city long enough to know better than to leave my bike unattended. I had had one stolen through my neglect once before. As I approached, the other woman turned to face me full on and I realized that she must be the Professor, Demi More-Terrafirma, whom Mona had mentioned before. She looked just like the actress Demi Moore.

Before Mona could introduce us, the other woman said, "Yes, I am the Professor and *she* looks like *me* not the other way around." I should know by now that all of Mona's crew possesses the same strange powers that she does. And, I should be used to it, but sometimes it can still throw me, especially when I am distracted. I have described how attractive Mona is and I was pretty much inured to it, but that still didn't prepare me for coming face to face with someone who was a dead ringer for a woman whom I had had a crush on for years.

I stammered and fumbled my way through an apology as Mona snickered behind her hand and Demi smiled sweetly. Mona ended my torture by saying, "Michael, let me formally introduce you to Professor Demi More-Terrafirma. Demi, Michael."

I didn't know weather to shake her hand, kiss it, or simply drop to my knees in supplication. She laughed and extended her hand in a way that I could clearly read as meaning that a simple handshake would be appropriate. As we shook hands I stammered, "Nice to meet you. Really, very nice to me you. It's a thrill to meet you." I sounded like a Tourettes patient, but without the cussing. And, to make matters more embarrassing, if that was possible, I still had her hand in a death grip. It was like rigor mortis had set in and I couldn't let go. I had seemingly lost control over some of my basic motor skills.

Mona didn't help matters by making no pretense to hide her amusement any longer. She was laughing out loud. Demi, no doubt used to this kind of reaction from full-grown men, was very gracious. She maintained her sweet smile, although I could see it was becoming more forced the longer I clutched her hand. I finally regained some semblance of control over my extremities (my feet had become rooted in place as well) and was able to release her from my grasp.

"Aren't you sweet?" she said. "But, you really must remember that I am not she. We are merely similar in appearance." *That you are.*

"Yes," I responded. "I will try to keep that in mind." Knowing that I would be unable to do so. The resemblance was just too remarkable and the duration of my crush too long to be able to turn off my mind to it that easily.

"Just try and not let it become too big of a distraction in any case", Mona said. I smirked at her and turned my gaze back to Demi. I figured Mona was just jealous because she wasn't the center of attention this time. She laughed out loud and said, "Please, Michael. Oh, and we will discuss the other matter when Demi has finished her lecture." She had obviously picked up on my emotional residue from my fall-out with Lily. I nodded my acknowledgement to Mona.

"What are we going to cover first today?" I asked, desperate to change the topic. I arranged my pack and readied my tape recorder.

"Demi is going to explain what destruction is befalling the land and the consequences of this destruction," Mona answered. She turned to Demi and said, "It's all yours, Professor."

Demi nodded at Mona then shifted her gaze to me and said, "Shall we get comfortable?"

I'm not sure what look came over my face at this point, but it must have been an extremely goofy one judging by their reaction—which was to laugh hysterically.

"Allow me to rephrase that," Demi said, after managing to catch her breath. "Why don't we all take a seat?"

We each picked out a patch of ground at the edge of the garden and 'got comfortable'. There was a small group of people working on one of the garden plots, but they were far enough away from where we sat as not to be a distraction. They were busy weeding and seeding their plot. They looked to be very content.

"I would like to begin," Demi said. "By explaining what the two main problems are that we face today concerning land degradation: soil erosion and desertification."

"We talked briefly about both of those in the discussion we had on the rainforests," I said.

"Yes," Mona said. "This does pretty much bring us full circle. It will be necessary to re-hash some of what we have already discussed—particularly about the rainforests, species extinction, and the interconnectedness of all life—in order to fully discuss the issues of land destruction and its related topics."

"I wasn't complaining. I was merely letting Demi know...can I call you Demi?" She nodded her assent. "I was just letting Demi know that I wasn't totally ignorant on the topic."

Mona gave me a 'yeah, sure' look which I responded to by sticking my tongue out at her from behind my hand. I didn't want Demi to see that—not that I was trying to impress her or anything. Mona rolled her eyes and said, "Are you going to be able to concentrate, Michael? I mean, with all you have on your mind." A not-so-subtle reminder of the problems I already had brewing with Lily.

"Yes, I will be able to concentrate, Mona. Thank you for your concern," I growled.

"That's good," she said and signaled for Demi to continue.

"I will discuss soil erosion first," Demi began. "Because it is what leads to desertification. In your country, the main cause of soil erosion is over-farming. Trees and hedges are ripped from the ground so as not to damage tractors or other farming equipment. Then, the land is ploughed too often and too deeply and fertilized too heavily because the soil has lost a great deal of its minerals and nutrients through overuse. Eventually the wind carries away the now dry, exposed surface soil. What the winds don't carry off, rain washes away because there are no longer any root systems to hold the soil in place. The U.S. has lost one-third of its arable land in the past 200 years and almost all of its original grassland over the same period. You are losing topsoil at the rate of 5 billion tons per year in the U.S. alone and another 21 billion tons worldwide."

"Twenty-one billion tons? That's an incredible amount," I said. *Now there's a comment on the obvious, Einstein.* Mona must have been having the same thought judging by the way that she rolled her eyes at me again. I still think she was jealous. I gave her a smirk and returned my attention to Demi.

"Other countries," Demi continued. "Most notably South and Central America, are destroying their land by chopping down the rainforests. What happens in this case is the poor people of these regions cut down the forests to create pastureland to raise cattle to sell to the Western cultures who have made beef a main staple of their diet. Or, they use the land to produce food crops because corporations have driven them from their ancestral lands in this land-grabbing environment.

"This is very shortsighted because the soil in rainforests is not designed for either of these uses. In the rainforest, as a result of their tropical climate, decomposition of waste matter is so rapid that the nutrients normally found in soil are retained in plants and animals instead. The trees are chopped down and the ground is slashed and burned, which also kills off or chases away the animal life. So, the soil is left almost barren and whether it's used for pastureland or to grow crops, it is quickly depleted. After only a few seasons, crops fail or the pastureland is denuded, so the people and the corporations move to another area and cut down more trees starting the whole vicious cycle over again."

"Seems like the natives would know better," I said. *I wonder if she's married?*

"They do," Mona said. "But, they are left with little choice. It's the corporations and the consumers demanding the products who are to blame."

"The consequence of all of this rainforest destruction and soil erosion," Demi said. "Is the desertification of the land. Presently, the deserts of the world are growing at a rate of 5,800 square miles annually. Almost 50 million acres of once fertile cropland is rendered unproductive yearly. And, slashing and burning the forests, poor crop production practices, and over-grazing are not the only causes of desertification. It is also exacerbated by poor irrigation schemes that lead to salination, which is the salting of the water supply, and by land getting paved over to make homes, cities, roads, and such. In all, fully two thirds of the world's agricultural lands have suffered from soil degradation in just the past 50 years and one third of the world's original forests have been converted to agricultural use to make up for the loss."

"Unbelievable," I mumbled. *As are the color of your eyes.*

"The danger in all of this is that as the human population continues to grow, you have less and less land with which to feed it. It takes the planet 400 to 500 years to build one foot of topsoil. You are losing topsoil at the rate of one inch every 16 years. You do the math. This whole process of forest destruction, soil degradation, and desertification is reducing the earth's ability to support livestock and crop production.

"An offshoot of the massive amount of livestock production that is taking place is the amount of animal waste it creates. In the U.S. alone, livestock excrement is produced at the rate of 800 million pounds per hour. And the worldwide figure for methane being produced just by cattle is 100 million tons per year. The methane ends up in the atmosphere worsening the greenhouse effect and the solid waste ends up in rivers and streams affecting the water supply."

"You mean to tell me that cow manure goes into the water supply?" I exclaimed.

"Yes. And, we also have the matter of human waste. Annual production of municipal solid waste is 217 million tons. You have one particular dump in New York that is the highest point on the Eastern Seaboard south of Maine. It is piled so high with trash that it has its own airplane-warning beacon. You have over 300 cities worldwide of one million people or more producing massive amounts of waste every day and you are adding another one million people to the planet every four to five days. All told, in America alone, each person produces an average of 25 pounds of trash per week when commercial and residential waste is figured in. Add in non-hazardous industrial waste and you get another 7.6 billion tons. Try to bury it, you harm the land; try to burn it, you harm the air; try to dump it in the oceans and rivers, you harm the water."

Mona interrupted at this point. "We have covered air and water pollution showing how you are in fact ruining the air you breathe and the water you drink. We will wrap up the topic of land degradation by showing how this has an adverse effect on the food you eat. And, conversely, how the food you eat has an adverse effect on the land."

"Fine by me," I said. *Wrap up the topic! Already! Does that mean Demi will be leaving soon?*

"The danger in using up your natural resources at such a rate," Mona said, while giving me a look like 'Grow up, Michael'. "Is that they take so long to replace, as shown by the example of topsoil. If in fact they can be replaced at all. There are essentially four types of capital: 1) human: labor, intellectual, cultural; 2) financial: cash and investments; 3) manufactured:

infrastructure, tools, machines; and 4) natural: resources such as coal, oil, and gas; the ecosystems by which I am referring to the oceans, forests, savannas, wet lands, estuaries, coral reefs, and tundra; and animals. Humans are easily replaced as witnessed by your population explosion. Nine months, instant capital. Financial, just print more money…or start up an Internet business. Manufacturing, no problem, build another factory, staff it from your huge labor pool, produce things. But natural resources, not so easy. Once they are compromised, polluted, or outright destroyed, the replacement time is enormous. In the case of coal, gas, and oil we're talking millions of years; rainforests, oceans, and tundra you're looking at centuries; estuaries, wet lands, and coral reefs take decades. Animals, millions of years, if ever. Demi, would you care to pick up with food issues now that I have mentioned animals?"

"Absolutely," she said. "I take it that you have touched on how past civilizations self-destructed due to leveling their forests to feed their ever increasing populations which eventually led to depletion of their lands, correct?"

"Yes," I said. *She is so hot. Okay, focus knucklehead.*

"Good," she continued, a slight smile playing at the corners of her mouth. "Now I will explain how you are repeating history, but with a subtle difference. Civilizations such as the Sumerians, Greeks, and Romans used the land mainly for agricultural purposes, not to produce meat products. They supplemented their diet with meat; it was not the main food source. Your culture on the other hand, uses the newly acquired land as grazing land for livestock instead of to grow agricultural products. You make your main course meat and supplement your diet with grains, vegetables, fruits, and such—that is, what grains are left over after you feed your livestock."

"I'm not sure I'm following you here," I said. *And, I would follow you anywhere.* "Are you saying that we are using food that could be going to feed the starving children to fatten up animals instead?"

"That's exactly what I am saying. Upwards of 85% of the top-soil loss is due to livestock production. In your country alone livestock consumes 90% of the soy and 80% of the corn that is grown. Of the grain that you export to those starving countries, livestock consumes 66%. And this when a child dies every three seconds from malnutrition."

"I never realized," I said. *I guess I should focus.*

"Not many of you do," Mona said. "That's why this book is so important. Maybe if your people hear it directly from Mother Nature and her friends, they will be more open to what is not only taking place in their own back yard, but also around the world."

"The point we are trying to make here," Demi said. "And the shame of it is that you don't need all of this livestock—cattle, pigs, chickens, and turkeys in the form of meat and their by-products—to maintain a healthy diet. Especially in light of the costs. And, while we are talking about animals for food, it should be noted here that you have also depleted 70% of your fish stocks worldwide. Besides the deaths caused by starvation that have been repeatedly mentioned, there are myriad other health issues directly attributable to livestock production and consumption.

"Livestock are fed antibiotics, which results in the development of antibiotic-resistant bacteria that is passed on during consumption of the animal product or merely by contact in some cases. Staph infections resistant to penicillin have risen from 13% in 1960 to over 90% today due to this procedure. One third of chickens tested today test positive for salmonella bacteria. It has gotten so bad that when polled, three quarters of the poultry inspectors said they refuse to eat poultry."

"And that poultry still makes it to market?" I asked.

"Oh, yes," Demi answered.

"My gosh," I said. "That would be like the quality control inspectors at my place of employment refusing to drive the cars that are produced there."

"A very apt analogy," Mona said. "In addition, the pesticide used on the grain that is fed to the livestock is turning up in humans at an alarming rate. As I mentioned in the chapter on air pollution, in mothers classified as meat eaters, when their breast milk was tested, it was shown that 99% of them had significant levels of DDT in their milk. That is a substance that was banned over 30 years ago as a lethal toxin, yet it is still showing up today. Chloraphenicol, iprenidazole, and cabadox—lethal toxin all—are more examples of pesticides that are used to treat grain that is fed to livestock and that are found in humans. The irony is that while pesticide use is up 3,000% since 1946 crops in general have decreased in yield by 20%."

"And, with corn in particular," Demi said. "The average amount of pesticide applied per acre has risen 100,000% the last 40 years while corn production is down 400% over the same period. Now, does that make any sense?"

"Uh, no, not at all," I replied.

"Fully 68% of your diseases are diet related," Demi continued. "Some other health problems greatly exacerbated or directly caused by this chemically laden, high fat diet includes arthritis, diabetes, kidney disease, osteoporosis, strokes, impotence." She gave me a rueful glance and I sat up more erect to indicate that I didn't suffer from that particular problem. She and Mona chuckled to each other before Demi continued. "And, obesity."

"Pizza and chocolate and burgers, oh my," I said.

"Yes," Mona said. "I mentioned the obesity statistics earlier for the United States. In England, obesity has doubled in the past ten years to 16%. And, a little historical footnote: in the 1970's your airplane manufacturers had to widen the seats in their planes because the average passenger had grown too fat to fit in the existing ones."

"You've got to be kidding," I said.

"Nope. For goodness sakes, Michael, fat makes up fully 40% of the calories consumed by the average American. Your medical professionals say that figure should ideally be between 15% to 20%."

"Wow," I said.

"Yeah, wow," Demi said. "Then there is heart disease, the number one killer of humans in the U.S. This when a meat-free diet reduces the risk of heart attack by 90% and your country spends $135 billion annually to treat cardiovascular disease. Lastly, 40% of your cancer cases are diet related with the big three being colon, breast, and prostrate cancers, respectively."

"I can't believe we do this to ourselves," I said. "It's like we are suicidal."

"Funny you should mention that," Mona said. "Since the 1940's, suicide rates have risen in every age group, as have the rates of major depression and alcoholism. And, when Western Culture is introduced in other countries, such as Japan, their rates go up in these areas, too."

"I guess there's something to be said for a simpler lifestyle."

"You think?"

"Like all other land mammals," Demi said. "Humans depend on four primary resources for survival: breathable air, drinkable water, arable soil to grow food, and a diverse, supportive biota. This one facet of your cultural make-up, this love affair with meat and poultry products, affects all of these areas in a negative way. We have already talked about the amount of arable soil that you are losing so I will skip that and pick up with breathable air. You have heard how critical it is to monitor carbon dioxide and methane emissions. It takes the burning of 200 gallons of fossil fuel, producing two tons of carbon dioxide emissions to produce the beef consumed by an average American family of four in a year. These same cattle emit over 100 million tons of methane into the atmosphere. Currently 40,000 square miles of rainforests in South and Central America are destroyed annually to produce the 300 million pounds of beef that the U.S. imports from those regions. You have already heard the consequences of rainforest destruction so I won't rehash those here either. The irony is that it is generally twice as profitable to harvest produce as it is to raise cattle, but the Western cultures demand meat."

"So, then," I asked. "The Law of Supply and Demand in respect to the consumption of meat products is actually a lose-lose proposition?"

"Yes," answered Mona. She motioned for me to look at the group working on the garden plot. "Imagine how many people could be fed and how much money could be made on the produce harvested from that one plot opposed to the totals generated by the one or two cows that could be fed from the same amount of land."

"You're right," I said.

"Unfortunately farming has its problems, too," Mona said.

"Yes, it does," agreed Demi. "And as for drinkable water, over half of the pollution in wells and surface streams world-wide is a result of agricultural production and in the U.S., the amount of water pollution directly attributable to agriculture—including soil runoff, pesticides, and manure—is greater than all municipal and industrial sources combined. On another front, livestock production accounts for over half of all the water consumed for all purposes in this country. The monetary cost of livestock production in the U.S. is almost $1 billion per year in the form of taxpayer subsidized irrigation practices. Globally, the World Bank has spent almost $2 billion financing livestock production in the last 30 years.

"Last, but not least, is the matter of diverse, supportive biota. Diversity is so important because it ensures survival. The more diverse the life in an ecosystem, whether we are talking about plant, animal, or cultural, the better the chance for survival of the various species when the system is faced with catastrophic events. Nature was designed that way. Your major extinction events over time have shown that fact to be true. I'm sure you have heard the saying 'variety is the spice of life'?"

"Sure," I said. *Would you care to spice up mine?* She and Mona both gave me a look. Whoops.

"Well," Demi continued. "Your culture has narrowed its choices to the red zone level. Out of the 7,000 or so crops that have been raised for food throughout human history, you rely on about 20 today. And, out of those 20 crops, three—maize, rice, and wheat—account for over 50% of the total. If one or two of those fail due to poor soil, drought, or other natural (or unnatural) disasters, you are going to be in big trouble."

"And, since Demi mentioned droughts," Mona said. "It should be noted here that droughts in the U.S. Southeast over the past year or so are the worst in the last 100 years. This might be due to the fact that this past spring (2000) was also the hottest and driest in the last 85 years. And, you wonder why parts of Georgia and Louisiana look like the Arizona desert."

"Coincidence," I said. No one laughed. Mona motioned for Demi to continue.

"One of the reasons that you have become so specialized in the choice of crops you produce is to feed livestock," Demi said. "You use 30 times the amount of cropland for this purpose as you do to produce grains, fruits, and vegetables for human consumption. On the same acre of land that you use for cattle to produce 250 pounds of edible product, you could produce 40,000 pounds of potatoes. Again, while a child dies every three seconds from hunger. That is nothing if not a spiritual abomination."

"I agree," I said. "It is an abomination. I also agree that we don't need meat to survive, that we can subsist on a vegetarian diet. But, why then do some doctors, school administrators, etc. push meat diets?"

"For the same reason," Mona said. "That you continue to destroy the environment in spite of the evidence that it is harmful and dangerous to do so—money. The people who push meat products in schools are sponsored by the meat and dairy industries. Just like any of the other harmful products such as tobacco, firearms, and gas guzzling SUVs, the industries and their lobbyists contribute heavily to political campaigns and thus have a strong voice in policy making decisions, legislation, deregulation and the like. Or, the lack of it. I mean, it has been proven beyond a shadow of a doubt that tobacco products kill tens of thousands of people yearly but until very recently they were still being advertised as a cool thing to do and pushed on the youth of your world. Your country has by far the highest rate of death by gunshot of any industrialized nation, yet the laws and regulations in place to keep people from obtaining them are weak and, in many cases, virtually unenforceable. And, when somebody tries to get tougher legislation passed, the people against gun control hide behind the 2nd Amendment of your Constitution when in fact it says, 'A well-regulated militia'. Well-regulated means legislated, controlled. And, it's a fact that an overwhelming majority of your automobiles pollute the environment, yet they continue to be mass-produced."

"This is a capitalistic society," I said. "The tobacco, gun, and auto industries, like the meat industry, are just operating in accordance with the law of supply and demand."

"Exactly. As do most all of your other industries including those that produce or supply fur coats, veal, and leather products. That's why we have to continue to educate the public on matters such as these. With education comes awareness, with awareness comes enlightenment, with enlightenment comes a shift in consciousness. And, nothing short of a shift in consciousness will bring about the needed changes. Did you think about the question that I put to you after our last meeting?" She explained to Demi what she was referring to.

"Yes, I did," I answered.

"And?"

"Well," I began. "What I've read about the collective consciousness is that everything is energy, including thoughts. When a thought occurs, it becomes a part of this 'consciousness'. Every thought that has ever been…thought is out there in that stream. That's why there are cases of people coming up with the same inventions or solutions to problems half way around the world from each other at the same time. It is also why entire cultures can be influenced to think and act the same way—Hitler and Nazi Germany, of course, being the prime example. But, it can also have a positive effect as in the issue of smoking. As was mentioned, smoking was very socially acceptable until recently. A committed minority engaged in a battle against the tobacco industry to get the message out concerning how bad tobacco products are for a person's health and things gradually changed. Now smoking is banned in most public places. Big Tobacco has been exposed as the evil entity that it is, and the legal and financial settlements have begun to take a serious toll on the industry.

I'm sure that it is this type of action that you are hoping to create in regards to environmental issues."

"Yes, it is," Mona said, smiling. "You really did spend some time thinking about this. Thank you."

"You're welcome." I have to admit; it felt nice to contribute in some small way. Mona asked Demi to take up the slack.

"Along with education," Demi said. "You also need to pro-vide solutions, or at the very least, alternatives. Its not enough to just say 'this is wrong, don't do it'. You need also to say 'Here is a different way to do it. The way we are living now is not work-ing, how about we try this way?' Notice I didn't say better way, I said different way. One of the main problems with cultures such as yours is the assumption that your way is better. If that were true, then how do you explain how the environment got to be so ravaged? What's so better about your way?"

"You have a point, I guess," I said. "But, what is a better, er, I mean a different way?"

"We will discuss that at our next meeting," Mona chimed in. "It is time to finish up. We have kept Demi long enough." *No we haven't!*

"No problem," Demi said. "I have enjoyed the opportunity to be of assistance. This is a vitally important project that you have undertaken."

"Thank you for all of your insights," Mona said.

"Yes," I said. "Thank you very much for coming and explain-ing these things to me. I really appreciate it. It has been a real pleasure listening to you. You make things so easy to under-stand. You have been the best speaker by far." *Oh, God, am I a dweeb or what? I sound like a love struck teenager with zero social skills. I'm turning into Spats (the Spaceman) without the brains.*

"Well, thank you, Michael," Demi said, suppressing her laugh-ter, "For all of your kind words."

"Maybe Demi will return to help us with our discussion of alternatives and solutions," Mona suggested, also suppressing laughter.

"I would be delighted," Demi responded.

"That would be great," I concurred.

"But for now let's call it a day," Mona said.

Demi got to her feet and made ready for her departure. "I must be going now." Mona and I stood up, also. "Mona, if you wish my assistance again, just let me know where and when." Mona nodded her assent. "Michael, in the event that I don't return, good luck with your project. Don't give up. There is too much at stake."

"I'll do my best," I said. "Thanks, again, Demi."

"You are quite welcome. Goodbye you two," she said as she walked away.

Mona and I both said a last goodbye. Then Mona turned to me and said, "Are you ready to discuss what happened with Lily?" I reluctantly turned to look at her. I would have preferred to watch Demi until the instant that she was…gone. I assumed that she could tunnel like the rest of her friends. Crushes were certainly strange things. Then, again, most all of my dealings with respect to women were strange—my past a testimony to that observation. Which led back to my present situation with Lily. Mona listened patiently until I concluded my replay of the previous night's events and asked for her feedback.

"Honesty is the best policy," was her response.

"Yes, Mona, thank you. But, that doesn't offer much in the how-do-I-make-it-right category."

"Well," Mona said. "If you had been straight with her from the start you wouldn't be in this mess. Not that I'm going to say I told you so." *No, of course not.*

"Look," I asked. "Are you going to help me or just continue to quote bumper stickers?"

"A bit of both, I think. I have to get my kicks where I can, Michael," she said, throwing my own philosophy back at me. *She is merciless. But, I guess I deserve it.* "Yes, you do," Mona said, but with a smile this time. "My suggestion is that you give her some space. She will let you know when she's ready to talk."

"I was afraid you would say something like that," I said, bummed out. I don't do patience very well.

"I strongly advise," Mona said. "That you ask Lily to accompany you to our meeting next Saturday."

"What if she's not talking to me by then?" I asked.

"Oh, I suspect that she will be," Mona replied. She was acting even more mysterious than usual. "I also suspect that she will at first refuse your invitation, but that if you are persistent, she will relent and come along."

I could only imagine how she 'suspected' all of that, but let it go. I did ask, "What purpose will it serve to bring her along?" I was mainly thinking of Lily's reaction when she saw how truly beautiful Mona was (ditto Demi, if she were there). I could see nothing good coming out of that.

"Trust me, Michael. Just make the offer and be insistent, but humbly so. I'll take care of the rest."

"All right. I'll do my best to get her to come along…against my better judgment, I might add."

"Like your judgment has been good so far."

"Ouch."

"The truth hurts sometimes," she said. *Great, more bumper stickers.*

"Where and when are we going to meet next Saturday?" I asked as I started gathering my things to leave.

"How about 9:00 am at your Museum of Natural History?"

"Very appropriate, Mona," I said as I unlocked my bike and mounted up.

"I thought so," she said. "Until then."

"Until then," I replied. I took off for home, my mind racing with the day's events.

INTERLUDE

TRUSTING MONA

Mona had been correct of course. Lily maintained her communications blackout until Thursday evening when she called to say we needed to talk. Of course, she didn't want to talk right then because she wasn't ready to, but would be ready to the next night. The reasoning behind that escaped me—that she wasn't ready to talk Thursday, but knew she would be Friday—nevertheless, I chose not to pursue that with her. I imagine the female readers understand that thought process, but it eludes me entirely. I admit to being at a total loss when it comes to interpreting the female mind and I'm also sure that it's entirely my fault. Lily asked if Friday evening would be acceptable to me, that was if I didn't have plans already...a not-so-subtle dig, I thought. I didn't have plans and told her so and agreed to be at her house at 7:30.

Aside from the dig about me possibly having plans for Friday night, the conversation the previous evening had been civilized. That gave me hope going into this meeting. When will I ever learn?

"You want me to come with you on your next date?" Lily shouted. We were in our usual places—me behind the couch, her pacing the room, weapon (the unused paperweight from our last meeting) in hand.

"It's not a date," I said. "My meetings with Mona have never been dates. We meet, she and sometimes one of her friends, tell me about air pollution or the rainforest or something else to do with the environment, then we go our separate ways."

"So, you've already met her friends." *The things she chooses to focus on.* "If memory serves, we were seeing each other a few months before we met each other's friends."

"One more time. Mona and I are not seeing each other. She is a married woman who is sharing information with me in an effort to halt the destruction of the environment. I am seeing you." She raised an eyebrow to let me know that the chad were still being counted on that issue.

"You are unbelievable," she said. "You see this woman behind my back, lie to me about it, and now that you're busted, want me to chaperone your next date." *And, I'm the one who is unbelievable.*

"You wouldn't be chaperoning our date. You would be meeting her and taking part in our discussion. Lean Gene might even 'tune in'. And, some of her friends will probably show up, too. You've got to see them tunnel."

"Tunnel?"

"Yeah, they just materialize, but they call it tunneling." By the look on her face, I guessed that I had said too much as usual. "I know it sounds weird."

"Now there's a world class understatement," Lily said as she juggled the paperweight in her hand.

"That's the same reaction I had when Mona started laying this stuff on me," I said, hoping to bond by showing some common ground.

"*Laying* this stuff on you?" So much for bonding.

"Okay, poor choice of words."

"A Freudian slip, perhaps?"

"No just…no, not at all. Please, just keep an open mind. Again, you have had some experiences in your life that should allow you to at least entertain the possibility of this being true."

"It would be easier to believe if you weren't telling it to me because you were caught in a lie."

"You really need to get past that," I said. Open mouth, insert foot.

"Yeah? I do, huh? You have some nerve. I 'really' need to get past this!" *Wow, she actually made the quotation mark signs with her hands like Mona does. I don't think I'll point that out now, though.*

"You're right," I said. "I'm sorry. I'm just frustrated, that's all."

"Ah, you're frustrated. Gee, isn't that just a shame." She could really grind my gears when she put her mind to it. I gave up trying to convince her of anything at this point. Maybe if I could just get her to come with me Mona would have better luck. Maybe use her necklace or do some kind of Spock-like mind meld. Bottom line, she couldn't do any worse than I was doing.

"Lily," I said using my most persuasive, ingratiating voice. "Please just come with me tomorrow and you'll see that this has all been perfectly innocent."

"You call lying to me perfectly innocent?" *Focus, Lily, focus.*

"I wasn't referring to the lying. But, really, after you meet Mona and her friends, you will see that this has all been a huge mistake."

"Excuse me?"

"On my part. A huge mistake on my part."

"Any other surprises?" I considered telling her about Demi in the event that she showed up. Lily knew about my crush on Demi Moore. Hell, I was on my second video of Ghost having worn out the first. I couldn't even walk by a piece of pottery without getting excited. But, I reasoned that at this point it would do more harm than good. I kept flashing on Mona telling me to just get Lily to come and she would take care of the rest. I shook my head 'no' in answer to her question. She stopped pacing and put the paperweight back on her desk.

"So, will you come with me, please?" I asked.

"I'll think about it," she answered.

"That's all I ask." I knew her well enough not to push her any further and, that 'I'll think about it' usually meant 'yes' in Lily-speak. I got up to leave. "I'll call you in the morning, okay?"

"Fine." She didn't make any move to walk me to the door so I let myself out.

"I'll talk to you tomorrow, then. Good night, Lily."

"Good night, Michael" I heard her say softly as I was shutting the door.

I somehow managed to get a good night's sleep. At least as good as I can get with an insomniac cat sharing the apartment with me. Nikita woke me per usual at 5:00 am, but I was able to fight her off and go back to sleep until 6:30-during which time I had a very erotic dream about Demi. I just hoped that if she showed up today that she wouldn't be able to intuit that. I could just see her and Mona huddled together giggling at me with Lily looking on.

I called Lily at 8:00 and she made me grovel for a few minutes before reluctantly agreeing to go with me to my meeting. I picked her up at 8:45 and, after a few failed attempts on my part to jump-start a conversation, we drove in silence to the Museum. I was a nervous wreck wondering about whom was going to be there and how Lily was going to react to how beautiful Mona really was (and the whole 'Demi' thing, of course). But, it was out of my hands. I had to trust Mona.

CHAPTER 8

SOLUTIONS

When we pulled into the parking area, Mona was nowhere to be seen. *Maybe Lily was going to see an example of tunneling. Great. Freak her out from the start.* My trust in Mona began to wane as my anxiety grew. I parked the Jeep and Lily and I walked towards the entrance. Then, Mona appeared…coming out the door, thankfully. But, every blessing has its curse, I guess. She looked especially lovely today—once again sporting her blonde/blue-eyed facade—and this had not escaped Lily's notice. I could tell by her sharp intake of breath and the muttered 'my, God'. By now I was on the verge of a full-blown panic attack. Mona walked to within a few steps of us before smiling radiantly and saying, "Good morning, Michael."

"Good morning, Mona," I said. *Here we go, battle stations.*

She quickly turned her radiant smile on Lily and said, "And, you must be Lily. I'm Mother Nature, but please, call me Mona. I'm so glad to finally meet you." Her manner was so disarming and sincere that I could feel the tension that was coming off Lily in waves begin to dissipate.

"Thank you for inviting me," Lily said. "Or for convincing Michael to invite me I should say." It was obvious that the dig was aimed at me, not Mona.

Mona smiled in acknowledgement. She turned to me and said, "Michael, why don't you go on inside. Perhaps visit the gift shop. I would like to have a moment with Lily." She glanced at Lily and said, "If you don't object, of course."

"Not at all," Lily replied.

I wasn't at all hip to this, but I didn't see that I had a choice, either. So, I mumbled something about getting some coffee and went inside. I watched from the gift shop window as Mona and Lily chatted in the parking lot. And, that's exactly what it looked like, too. A couple of people chatting. Not having a serious discussion. Not arguing. Just chatting. Like it was the most natural thing in the world for Lily to be having a friendly conversation with the woman she suspected of fooling around with her boyfriend. Assuming I still was her boyfriend, of course.

To this day, I don't know what was said between them. Lily has refused to tell me and I haven't been able to intuit it despite my increasing ability with that gift. But, whatever it was, after fifteen agonizing minutes they walked arm in arm into the Museum like life long friends—go figure.

At Mona's suggestion we took a walk through the museum. It was fairly deserted. Besides the fact that it was barely nine o'clock on a Saturday morning, there were some special events taking place around town that weekend: a Jazz festival at Courthouse Square and the annual county fair at the Montgomery County Fairgrounds. Many of the people who would have normally been availing themselves of the many exhibits at the Museum of Natural History had been drawn to these and other outside events as it was also another beautiful spring day. Partly cloudy with a promise of temperatures in the mid-70's.

We walked along three abreast, Lily in the middle still arm in arm with Mona, the two of them chatting away about the exhibits, excluding me from their conversation. This was just as well as I was preoccupied with a conversation of my own—the one that happened to be taking place in my head featuring two fears that were personal favorites of mine: Jealousy, who was complaining about the instant bond that had occurred between Mona and Lily; and Worry, who was pondering the implications of said bond.

We continued on like this for a half-hour or so, until we came to a door marked 'Auditorium'. Mona opened the door and motioned us inside. The auditorium contained seating for about 150 people, but was currently unoccupied. I asked Mona why we were using such a large room. She said, "You never know who might show up." The twinkle in her eye sent shivers down my spine as I immediately thought of Demi putting in an appearance. I wasn't at all certain that Lily would bond with Demi as well as she had with Mona. This was followed closely by a flashback to my dream of earlier in the day. I noticed Mona blushing and knew that she had intuited this. Luckily Lily hadn't picked up on it. "Why don't we take a seat," Mona said. "And get started on some possible solutions to the planet's ills."

We made our way to the front of the room. There was a large stage that held a conference table with a dozen comfortable chairs around it, five on each side and one at each end. We climbed the four steps up to the stage. Lily and I took seats across from each other at one end of the table and Mona sat between us at the table's head. She looked at Lily and said, "I know that you haven't been privy to everything that Michael and I—along with my husband and a few friends—have discussed so far, but considering your background with respect to environmental issues I'm sure you won't have any trouble following along. If you do have any questions or comments, don't hesitate to voice them, okay?"

"Okay, thank you," Lily answered.

"You're quite welcome, dear. Now where to begin?" She stared off into space for a moment. *Oh, no, she is going to channel Lean Gene.* But, she merely refocused, shot me an amused look having intuited my fear, and then directed a statement my way. "As you well know, before one is amenable to looking for solutions to a problem, they have to first acknowledge that a problem exists. Hopefully, we will have convinced the reader of that by the time they reach this point in the book. If not, this final chapter becomes irrelevant. So, make sure that the chapters prior to this are convincing, okay?"

"Sure, no problem, Mona," I said. "I mean, since I'm not under any pressure or anything."

"You will do wonderfully, Michael. Anyway, once the fact that there is a problem has been accepted, the shift in consciousness that we've talked about can begin to take place and the search for solutions can begin. Your Einstein was quite correct when he said 'a problem couldn't be solved with the mindset that created it'. You have all the tools you need to start a revolution of consciousness. You merely need to begin the shift. One person at a time, putting a positive paradigm into the collective consciousness, thereby influencing others, who will in turn influence others; this is how it will happen. Before you know it, you will have enough people thinking the same way to bring about the necessary changes. These changes will happen not only at the local level, but also on a global scale. So, again, if we have done our job, the people who have read the material up to this point will have accepted the fact that changes need to be made in respect to the environment."

"That," I said. "Would be what we humans call a 'comment on the obvious', Mona." *Let's get a little humorous banter going here.*

"Really," she said. She exchanged a look with Lily. *So much for banter.* I didn't need to be able to intuit to know that I was in big trouble. Mona continued, "Anyway, I believe that we should begin our discussion of solutions by first addressing the most pressing issue, the problem that started the whole ecological chain reaction of destruction: the population explosion."

"That is going to be a very controversial topic," I said. "You have people with extreme prejudices in this area, two very diverse belief systems at work. In China, the policy is one child per family. And, it is strictly and brutally enforced. Then you have certain Western religions that pretty much promote large families, the 'be fruitful and multiply' school of thought. You risk alienating large segments of the world's population no matter which view you favor."

"I don't favor any view," Mona replied. "I merely told you what the problems were of uncontrolled population growth and the consequences that are very likely to result from it if they aren't corrected."

"Okay," I said. "For the sake of argument, what are your suggestions for dealing with this problem?"

"Simple. Quit having so many babies."

"Wow, that's brilliant," I said. "Why didn't I think of that? Any other gems of wisdom?"

"Emphasize positions that are pro-death."

"Come again?" I asked. Lily looked as shocked as I was.

"Pro-death. There needs to be compromise between the two camps. I'm proposing an anything-that-thins-the-herd movement."

I realized that Mona was saying this with tongue planted firmly in cheek. I could tell by the look on Lily's face that she had picked up on this, too. We decided to play along. I led off with, "Compromise, huh? Interesting."

"Yes, very," Lily chimed in. "Okay, in that vein, how about the left agrees to a repeal of all gun control legislation if the right agrees to legalize all illicit drugs?"

"And, the left gives in on the death penalty if the right quits trying to have Roe v Wade reversed," I said.

"How about the left quits trying to hold big business responsible for their actions and the right quits trying to hold individuals responsible for theirs?" Mona said with a smile.

"In other words," I said. "No revocation of corporate charters or fining businesses for polluting or looting the environment in order to maximize their profits with which to feed their families and those of their shareholders in exchange for rescinding the Three-strikes your down for life and mandatory/minimum sentencing statutes for individuals who steal to feed their families."

"Very good," Mona said. "Now, of course, we are being facetious…for the most part. I'm sure there are those who would get behind any—or all—of these suggestions, especially if they could be directed at specific segments of the population. But, I'm optimistically certain that most of our ideas would meet with stanch resistance by a majority of the public. So, I guess we had better stick with realistic methods and goals. Which takes me back to 'quit having so many babies'. By this I don't mean that you should employ the Draconian methods employed by the Chinese government. What may be more readily acceptable would be to lower the infant mortality rate in the poor areas of the world."

"But, wouldn't that be at odds with the desire to lower the population figure?" I asked.

"No. Remember we said earlier that the reason that people in the poorer areas have so many children, especially overseas where they depend on children to supplement the labor force, is because of high infant mortality rates. In order to ensure survival of a few offspring, they have a whole slew of them and invariably, because of your indomitable human spirit, more survive than was anticipated. With lower infant mortality rates they would only need to bear the amount of children that they could afford."

"How would this message be conveyed?" Lily asked.

"Through education; access to adequate medical care; family planning and birth control techniques, which would become unnecessary after a generation or so as the benefits of this lifestyle change began to show themselves; better farming methods; healthier dietary habits; laws denying access of corporations to the lands of indigenous peoples. And, this is only in the poor areas, especially the Third World countries, which I am talking about. If the U.S. would cut its birthrate by only .5 children per family, your population would stabilize and then begin to decline by 2050."

"Would .5 be enough?" Lily asked.

"Yes," Mona answered. "And hopefully, starting with such a low, non-threatening expectation, people would feel obligated to try it. Then they would begin to see the benefits and, as in the Third World, escalate the figure naturally. You can't imagine what a relief a stabilized population would be to the planet. But, again, it will take a whole shift in consciousness, a more altruistic paradigm. Ideally, people will begin to think long-term and see how their actions have an effect on everything and everyone around them. They will see that if maybe they content themselves with one or two children instead doing a Waltons impersonation, then their neighbors can have one or two, also. And, the children that they do have will have a planet to inhabit."

"You really think that this could happen, this shift in consciousness?" I asked.

"Absolutely," Mona replied. "But it better happen soon."

"Why is that?" I asked.

"Because your George W. Bush, in one of his first acts as President, cut off funding for the very education, family planning, and birth control we were just discussing to any and all countries that allow abortion. The result of this will undoubtedly be an increase in unwanted pregnancies, especially in the poorer Third World countries, which will raise the demand for abortions, the very procedure President Bush is trying to curtail by his decision. Obviously, this will also lead to an increase in the world's population, not really something that you need at the moment."

"Unbelievable," Lily said.

"Not to me," replied Mona. "But, let's get back to the positives, the shift in consciousness that I mentioned. Michael, you asked if I really thought it could happen, correct?"

"Correct," I concurred.

"My answer is yes, absolutely."

"How can you be so sure?" I asked.

"Well, let me ask you, have you ever gotten out of shape? Put on weight, perhaps?" You notice she didn't ask Lily.

"Sure," I responded. "That's one of the reasons I eat a vegan diet and exercise the way I do. So I don't have to go through the struggle of losing weight, again. I have learned that it is much easier to keep it off than to take it off, especially the older I get."

"Exactly. But, tell me, when you were overweight and started to diet and workout, what was the process like. Was it easy at first or difficult?"

"Difficult. I had to exert a great deal of discipline in my eating habits. Like not eating stuff that I had grown very fond of such as chocolate and ice cream and chips and dip. And, I had to force myself to work out when I didn't really want to. It was a pain—no pun intended."

"Sounds right," Mona said, smiling. "But, if it was so difficult to enact this lifestyle change, why did you do it?"

"I wanted to feel better," I answered. "And, I was beginning to worry about my health and the wear and tear on my body. And, not long after I started the process, I began to see the results. My pants weren't as tight around my waist. I slept more restfully. I felt better physically and I looked better. As I began to look better, people noticed and commented on it. That had the effect of me wanting to do even more. It was a boost to my ego and my self-esteem, which improved my emotional and psychological health. It created a positive domino effect." I paused to consider what I had just said. Then, it hit me. "Oh, Mona, you are good. I see where you are heading here. I see the connection."

"Your consciousness shifted to a healthier outlook," she stated.

"Yes. And, that is what you were getting at when you brought this up when Demi." I stopped abruptly. *Uh oh. Me and my big mouth.*

"Demi? Who is Demi?" Lily asked while skewering me with her eyes.

Mona jumped in to save me, "One of my crewmembers. You will meet her later." Mona's manner and tone of voice once again had a calming effect on Lily and she was able to let it go. *If only I could. 'You will meet her later.' Wonderful.* "We were discussing the effects that a shift in consciousness had on your outlook. Don't you believe that the same thing can happen on a larger scale? Say with a group of your peers? One member of the group starts a beneficial activity, the rest of the group see the positive results, get involved in it themselves, then their other friends, family members, co-workers notice, they get involved, etc., etc."

"Happens all the time," I said. "Word of mouth can be a very powerful tool."

"Yes, it can," Mona said. "Now, don't you think it's possible that the same thing can happen with the environmental movement on both the physical and consciousness levels?"

"I don't see why not," Lily added. "That's how most religions started and I heard on CNN recently that the environmental movement now has more followers than any religion in history."

"Exactly and that's a good start," Mona said. "But, more awareness is still needed. To achieve that you will need to be flexible, have a well thought out battle plan, and build a united global front. Your World Wide Web can come in handy there."

"We can do that," Lily said.

"I believe that you believe that I believe that," Mona said.

Lily looked at me, thoroughly confused.

"She talks that way sometimes," I explained. "Just roll with it." Lily nodded her acceptance.

"Sorry," Mona said, "Anyway, my hubby, if you remember, touched on the fact that your culture is into instant gratification."

"Yes, I remember something about that," I mumbled.

"That's another area that requires attention. You will need to change that mindset along with the cynical viewpoint, the fixation on irony that your culture is mired in. You will need to get the point across that if 'life sucks' don't get all negative and apathetic—take action, do something about it. Your Edward Abbey, a wonderful friend of the Earth said, 'Sentiment without action is the ruin of the soul'."

"I really admire his works," I said.

"Then honor his memory; raise a call to action."

"We will," Lily said.

"Good," Mona replied. "Then let's continue with this thread of instant gratification. Nature generally doesn't create or destroy instantaneously. It is usually a long, slow, methodical process. That has been one of the biggest problems in bringing attention to the plight of the environment. The degradation of the air, waters, and lands; the destruction of the rainforests, ecosystems, and habitats; species extinctions—all have been gradual occurrences for the most part. And, that has the potential to be a big problem when it comes to cleaning up the mess too, because it is going to take time to see the results. Some things will rebound more quickly than others will. When you quit slashing and burning the forests, you will begin to notice the difference fairly quickly, especially if you live in those regions. Just as you will if you change your dietary habits. Noticing cleaner air and water, the return of species populations in various ecosystems, rebuilding coral reefs—all are going to take a little longer. You need to emphasize this in any and all of the communications you have on this subject whether it is in this book, in meetings, broadcasts, magazines, or online chats."

"Broadcasts? Meetings?" I asked.

"Surely you don't think your responsibility ends when you publish the book do you?"

"I never really thought about it to be perfectly honest," I said.

"Well, you better start," Mona said. "You are going to have to be putting the word out about the book itself and about what is in the book. You will have to be one of the spokespersons."

"Oh, jeez. Couldn't I get someone else to do that?" I looked in Lily's direction.

"Don't try to stick Lily with this," Mona admonished. "She is going to be busy enough with her own projects."

"Yeah, Michael," Lily said. *Great, a double-team.*

"It's just that I don't think I have the temperament for it," I said. (Okay, so maybe it was more of a whine.)

"Certainly you do," Mona responded.

"I really don't think so, Mona." *The very thought of it gives me the willies.*

"Why, Michael?"

"Because I get too impatient, too easily angered when met by stubborn resistance or ignorance or outright closed mindedness. I become overly emotional and the message gets sacrificed at the expensive of my anger. I tend to get preachy and intolerant and I don't see how that will be helpful."

"He's telling the truth, Mona," Lily agreed. "He's all that...and more."

"Thanks for your support honey," I said.

"I'm merely agreeing with you, Michael. And, don't call me honey."

"Yes, Michael, don't trivialize Lily's position," Mona said. "It's demeaning."

"Sorry," I said. *Gee, this is big fun.*

"At any rate," Mona said. "It will present you with a chance to practice patience and tolerance. It will be simple once you get used to communicating always and in all circumstances from a place of love." I could see she was serious, no tongue in cheek here.

"Gee, that's all? Just communicate from a place of love always and in all circumstances. Well, heck, no problem then. Especially from such an empathetic, laid back, compassionate soul such as myself. Start booking personal appearances. Let the healing begin." *This talk of 'communicating from a place of love' will alienate more people than Bill Clinton at a GOP fundraiser.*

"I hear what you're 'saying', Michael. And, unfortunately, I agree. Talk of love would probably alienate a lot of people. And, what an incredible shame that is, what an indictment on your culture, that speaking of love would elicit so much discomfort and skepticism and fear from people. But, you don't have to verbalize it all the time. Be subtle and let your actions carry the message as well. When you are attacked, react with serenity. Face the anger and question it in a calm, rational manner. You will be surprised how that can diffuse a situation and open the channel for true dialogue to begin."

"You know I will try my best," I said while shaking my head doubtfully.

"I've seen glimpses of that capacity in you, Michael," Lily said.

"Really?"

"Yes, really." *Well, who would have thunk it?*

"As I said before," Mona said. "You will do wonderfully. I have total faith in you."

"Thanks for your vote of confidence. Both of you. I just wish you were able to be the spokesperson, Mona."

"I'll be there in spirit."

"I'm sure you will." *Knowing you, it will probably be in more ways than just spirit.*

"Now," Mona said. "We were talking about the population problem. The benefits of a smaller population will begin to show in every area of environmental concern. But, again, this is an area that is going to take time, patience, and discipline."

"And, maybe this is an area where I shouldn't speak of 'luv' at all." I said 'luv' as Isaac Hayes would and did a little eyebrow bob. Lily rolled her eyes, another thing she had in common with Mona.

"That wasn't the kind of 'luv' I was referring to," Mona said, doing a pretty decent Isaac Hayes herself.

"I'm just saying."

"Well, since you brought it up, that's where the discipline comes into play, along with the other issues I spoke of such as education, family planning, and contraception."

"Those are some hot-button issues," I said.

"I know," Mona replied. "Again, compromise, common sense, communicating from a place of love."

"If you say so," I responded.

"I do," Mona said.

"And, I agree," Lily said, getting in her two cents worth. *I'm out numbered, out gunned, and, quite possibly, out of my mind.* "Okay, Okay," I said, holding my hands up in the universal gesture of surrender.

"The next issue I would like to address is that of the rainforests," Mona said. "Some of the solutions that could dull the saw in this area are stiffer logging regulations along with banning road building in forests and citizen pressure on government to quit selling off public lands to the highest bidders."

"Or, to the biggest campaign contributors," Lily said. "Although they are a lot of times one and the same. Money talks."

"But, less and less people are privy to the conversation."

"Good one, Mona," I said.

"Thank you, Michael. Speaking of money, taking some of it away from subsidizing environmentally destructive practices and allocating it to research and technology for alternative energy sources and eco-friendly products would help, too. As would an offer of tax incentives to landowners and developers to preserve habitats threatened by sprawl. Other areas to explore are the feasibility of Green Taxes and taking away the ability of corporations to deduct legal expenses from their taxes when found guilty of environmental abuses. Having corporations pay to clean up *all* of the messes they make instead of shifting the financial burden to the taxpayers would offer a deterrent. As would leveling heavy, and I mean ouch-that-really-hurts heavy, not just slap-on-the-wrist fines; giving reimbursement to individuals, communities, and states that have been hurt by corporate practices; requiring companies to allocate a certain percentage of their profits and their research and development budgets to developing Eco-technology as a part of their fines and punishment packages; organizing more events like 'Buy Nothing Day' and 'Turn Off TV Week'. These are just a few examples of ways to combat environmentally unsound business practices."

"You are a regular Robin Hood, Mona."

"Opposed to a lot of your corporations who could be described as robbing hoods?"

"All right, Mona," Lily said.

"And, since corporations have the same legal status as individual citizens, let them suffer the same fate as an individual who breaks the law."

"Corporations have the same legal status as individual citizens?" I asked.

"Yes," Mona said. "Go figure."

"How is that possible?" Lily questioned.

"Maybe I should give you a brief lesson on the history of corporations in your country."

"No," I groaned and made my fingers into the sign of the cross as if to ward off vampires. "Not another history lesson."

"I'll make it as brief and painless as possible," Mona said, laughing.

"Very well," I said.

"But, as Lean Gene."

"You're going to love this," I said to Lily.

Mona got that same blank look on her face as before when she was about to channel her husband, then said in his Darth Vader voice, "One of the common misconceptions about your American Revolution is that it was solely against the British government and the taxation issue."

"Oh wow, that is so weird," Lily gasped, her eyes bugging out.

"Tell me about it," I said.

"But very cool, too."

"I'm glad you approve," Lean Gene said through Mona. "May I continue?"

"Certainly," Lily said.

"Thank you," he said. *He never treated me this nice.* "As I was saying," he continued. "While the taxation issue with the British government was one major factor in the Revolution, another was the response by the colonists to the price gouging carried on by British corporations. Your infamous Boston Tea Party was as much protest against the British East India Tea Company for the way that the company was charging exorbitant prices because of its monopoly power as it was against the British government.

"As a result, for the first one hundred or so years after your United States were formed corporations were kept under strict control. They were granted charters and regulated by the states. And, the charters could easily be revoked. Ironically, it was another war, your Civil War that changed all of this. Due to the war effort corporations grew steadily stronger financially, politically, and legally. The bigger corporations began buying smaller companies. They were also able to buy politicians and judges. Finally, in 1886 your Supreme Court ruled in Santa Clara County vs. Southern Pacific Railroad that private corporations were individual 'persons' under the U.S. Constitution and therefore entitled to protection under the Bill of Rights."

"That's unreal. Why don't we get that changed back to the way it was?" Lily asked.

"Exactly," Lean Gene responded. "Why don't you?"

"The way corporations have it today," I said. "That will be a long-term goal. I mean with the lax campaign financing laws, lobbyists running amok, etc."

"True, but it's still an issue that needs to be addressed," Lean Gene said. "In the meantime, you could at least push to have corporations punished as individuals. When a corporation is convicted of wrongdoing punish it. Forbid it government contracts; make the shareholders liable for the corporation's actions, both financially and legally. If a corporation is found guilty of pollution, collusion, etc. revoke its charter, force the sale of the company. Then, put the money into funds for clean up and to aid victims of the crimes. At present, it is profitable for businesses to pollute the environment because it is the easiest, most efficient way to handle the problem.

"Lastly, you have various and sundry international organizations to which your government belongs. The World Trade Organization, the International Monetary Fund, The World Bank, GATT, to name a few, that base their decisions on economic factors at the expense of the environment. Changes need to be made there, also."

There was a pause as Mona shook her head and 'came back'. "As you can see, my hubby has spent his time on the Internet wisely." Lily looked at me confused.

"He's an Internet junky," I explained to her. She nodded her understanding.

"He sounds very nice," she said to Mona.

"Thank you. He is," Mona responded. *Yeah, right. A real peach of a fellow.* "Now, to continue," Mona said after giving me her look, "There is much that can be done through grassroots organizations. At the local and individual levels is recycling. Metals, papers, plastics…anything that can be reused is worth recycling. Re-using paper bags or using canvas or string bags instead of the paper or plastic bags at the grocery. Enacting legislation banning unwanted junk mail and newspaper supplements. That would serve the same purpose as what is taking place on the Internet with regard to personal information. Your legislators are trying to enact laws that would forbid the use of Social Security Numbers, etc. without the permission of the individual. You could use the same concept to cut down drastically on this blanket advertising material. Make it illegal to distribute it without the permission of the distributee."

"That is a great idea," I said. "It kills me every time I discard about ten pounds of unwanted junk advertisements from the Sunday paper."

"Then along with that, you could organize grassroots campaigns to adopt parks, roads, forests, gardens, waterways, etc. not only to rid them of litter, but to watch for other environmental abuses and to replant trees, nurture regeneration of species of plants and animals, to generally care for. My hubby informs me that there are already programs like this on the internet where you can 'buy' a piece of a rainforest or save a certain amount of acreage just by clicking on certain Web sites that are funded by advertising."

"He's correct," I said. "One of the neatest gifts I received last Christmas was a certificate for an acre of rainforest that my sister-in-law adopted in my name."

"That is a wonderful gift. Another area to be exploited is Eco-tourism—trips to these endangered areas. This raises awareness of their plight and appreciation for their beauty and uniqueness. Publicity can serve the same purpose. But, it needs to be positive publicity. Not the negative, demeaning actions such as assaulting individuals by throwing tofu pies in their faces or dousing them with fake blood. These actions are counter-productive and serve only to alienate the very people they are trying to reach. Though I am sure the protester's hearts are in the right place and their cause is worthy that is not the way to go about it. They are not communicating from a place of love. Instead, they come off sounding shrill and irrational. Their message goes unheard amidst an atmosphere of anger. They get labeled as 'extremists' or 'radicals' or 'tree-huggers' and everyone else even remotely involved with environmental causes or animal rights gets stereotyped and lumped in with them."

"I hear you," I said. "I am sympathetic to their cause. I love animals. But, when I hear about those types of activities I get turned off. One of these organizations recently used a public figure's potentially fatal illness to promote their cause. It was sickening. It's the same principal as when I am having a discussion with someone and they start screaming and get abusive. They may be making very good points, but I am not hearing them because of the way they are communicating."

"My point exactly, but, again, enough negativity," Mona said. "Another avenue is to point out the positive economics of environmental products and services. Your world market for pollution control devices, waste management, recycling, and energy efficient products—commonly known as Eco-technology—is projected to reach $600 billion by the end of 2001. And, you need to learn to use your technology more wisely. Send more e-mail instead of using hard copy. Paper is 40% of the solid waste total. Recycling currently accounts for 43% of new paper products and you could easily raise that figure to 75%. Coerce the info-tech companies to recycle their products, especially the ones made from plastics such as computers, cell phones, printers, and fax machines. This would force them to produce simpler, cheaper products that would also be more eco-friendly and help with the oil price problems you currently face."

"My research says that is the wave of the future," Lily said.

"The future is now," Mona said. "That $600 billion figure represents more business than either your aerospace or chemical industries are projected to generate in the same time frame. And, speaking of chemicals, you need to stress the financial advantage of vaccines and other health remedies that are extracted from the rainforests and other ecosystems. We talked earlier of the enormity of untapped potential for medicinal products that might be available from the plants, bacteria, fungi, etc. of the different ecosystems. And, as any study on health costs will show you, prevention is always cheaper than treatment."

"Yes, I have seen some of those studies," I said. "Same principal as my diet and exercise program: it's easier to keep it off than to take it off."

"What is that saying of yours? An ounce of prevention is worth a pound of cure?"

"We humans do come up with some winners every once in a while."

"Yes, you do. Well, that pretty much takes care of the points I wanted to bring up about the rainforests. Just keep in mind, as with all of these solutions for the various areas we will discuss, we are only touching the tip of the proverbial iceberg. But, as you read about and investigate the topics we have raised, you will stumble across other actions you can take that will lead you to other actions you can take and so on."

"Very good," I said.

"Then let's move on to air pollution. Obviously the first step is to raise the issue of the need for stricter air pollution standards or at least improved enforcement of the ones already in place. I am talking about policing factories, power companies, and the auto industry in particular. You also need to push for the strengthening of the legislation that calls for reducing the levels of carbon dioxide, CFC's, methane and other 'greenhouse' gases in the atmosphere—especially in light of the recent actions of the Bush administration that we mentioned earlier. Leveling higher taxes on gasoline and other carbon based fuels needs to be looked into as does reducing the use of pesticides, fertilizers, and chemicals on crops, plants, gardens, parks, golf courses and such while at the same time looking into natural, organic alternatives. Also, the diverting of funding into developing more environmentally sound products and alternative energy sources such as hydrogen, wind, solar, geothermal, and ocean/tidal power and fuel cells, the so-called micro-power sources needs to be expanded. Over the past few years the wind turbine industry has grown faster than the personal computer market. Wind power usage is up 1000%."

"That's good news, no?" I asked.

"Of course, but there is still a long way to go. And, there is still a huge, untapped potential here. With more advances in the technology and infrastructure in this area, three states—Texas and North and South Dakota—could provide enough electrical power through wind turbines to provide the electrical needs of the entire United States."

"Get out of here," I said.

"Oh, it's true. What's more, the cost of wind power has gone down $100.00 per kilowatt-hour over the past ten years."

"Yeah," Lily said. "I have been making some phone calls and looking at writing a grant proposal to the state about putting windmills on a plot of land on the outskirts of the town where I live. It's not as expensive as one would think."

"Exactly," Mona said. "As soon as word gets around, it will get even cheaper because of competition. You also have the enormous untapped energy resource of the sun. Even though the sale of solar energy products has risen 15% over the last seven years while costs have declined by 256% over the same period, it is still the most wasted alternative energy source in the universe. If you could harness just 1% of the sunlight that reaches the planet on a daily basis it would provide you with more power than all of your other energy resources combined."

"That is a staggering statistic," I said. "But, what about the costs of developing these products? Wouldn't they be enormous, also?"

"Yes, but again, prevention is cheaper than correction. Over the past 25 years or so, pollution production costs have risen 8%, but the benefits of the resulting cleaner air meant a total cost *reduction* of 15%."

"That should get the attention of Big Business," I said.

"Yes, it should. But, there are many more benefits to developing micro-power sources."

"Like what?"

"First of all, being smaller in scale than power plants, the various forms of micro-power can be built more quickly and sited more efficiently. Second, they generally emit much lower amounts of pollution such as carbon dioxide and sulfur. Third, they are easier and cheaper to maintain and repair. Lastly, and most important for your society I'm certain, is that they are independent of fossil fuel price fluctuations because they rely on local fuels and spur community economic development. Any questions before I continue?"

"Yes," I said. "Earlier you mentioned hydrogen power. From what I have read the huge advantage of hydrogen power is that hydrogen gas burns clean. In other words, without polluting. And, its only waste product is water."

"That's true," Mona said.

"Then why haven't we developed this form of power more?"

"Because of the stranglehold that the big oil companies and their auxiliary industries have on your elected officials, your lawmakers. The answer to this again lies at the individual and local levels. You need to organize citizens to vote out the representatives who support unwanted business practices and environmentally harmful policies and programs and vote in officials who will truly represent you and your interests."

"Bad time to be talking about elections and voting," I said.

"Really," Lily agreed.

"But necessary nonetheless," Mona said. "Other areas to stress, besides those already mentioned would be further development of affordable, sustainable public transportation; more car pooling; and increasing the amount of home-based workers or workers who live close enough to their job to walk or bike.

"Further studies on topics such as the hydrogen power we just mentioned and sequestering carbon dioxide in deep ocean sites or in coral reefs would be beneficial. Mandating that your local, state, and federal governments use only environmentally friendly products and services, such as buying only fuel efficient vehicles for their employees, would set a wonderful precedent. Force legislators to repeal such bills as the one pushed into effect in 1995 by the Republican controlled Congress and reenacted each year without opposition from the Democrats, that bans raising the standards on auto fuel efficiency and is so restrictive that it contains provisions barring even the study of ways to improve efficiency."

"That's outrageous," I said.

"Isn't it though? But, you sit idly by and let it continue. I find it very ironic that one of the main issues in the most recent Presidential election of yours was education when in reality the folks running the country seem to want to keep you as uninformed as possible, in the dark as it were—no pun intended, of course." *Yeah, right.*

"I never thought of it like that," I said.

"My point exactly," Mona replied. "Now, let's move on to water pollution. Many of the solutions that we have already pointed out pertain also to water pollution. Solutions such as Green taxes, recycling, stronger legislation, and stricter enforcement of existing legislation—in this case the Clean Water Act of 1972—so this will be a short section so as not to be redundant."

"Makes sense to me," I said.

"One specific area that we can discuss here is the issue of acid rain," Mona said. "Obviously, cutting down on air pollution would have a very advantageous effect here. Another suggestion is to buffer lakes that have been affected by acid rain. That is treating them with natural acid to help offset the unnatural acid content and balance the pH levels. Monitoring wells, aquifers, lakes, rivers, and the oceans more closely for bacteria levels, acidity, sewage and the like is another area to explore.

"Individuals will want to look into investing in low-flush toilets and low-flow showers and faucets. Being diligent about fixing leaks and conserving water by such methods as using wastewater on plants, gardens, and lawns, and demanding that their municipalities do the same in regards to golf courses, parks, and playgrounds would be of great benefit.

"Another area to look at is in the improvement of irrigation practices. In the technical arena you could use more efficient sprinklers to apply water more evenly and cut evaporation and wind drift losses; in the managerial realm, implementing better scheduling to improve canal operations for timely deliveries so as to apply the water when it is most crucial to a crop's yield, along with better maintenance of canals and equipment would be beneficial; institutionally, you could instigate establishing water user organizations promoting more involvement of farmers, and collection of fees, establishing a legal framework for equitable water markets, and offering better training; and in the agronomic field you could foster a program of inter-cropping to maximize use of soil moisture, plant drought-tolerant crops in areas where water is scarce, and experiment with breeding water-efficient crop varieties.

"Along with this, a push to toughen up laws to disallow subsidies to mining companies would be nice. Currently there are one-half million abandoned mines, a lot of which are leaking toxic waste into lakes, rivers, and streams. These mines have also destroyed a great deal of land—which creates a natural segue into land topics. In this area, the banning of certain chemicals, fertilizers, and pesticides is crucial. Enacting legislation to control what goes into those substances would help, too. For example, at present you have no laws against using toxic waste or radioactive materials in fertilizer in the United States."

"That's insane," Lily said.

"I agree," Mona said. "Another area to explore is reducing solid waste through recycling. Burning it doesn't work because all that accomplishes is to change it to air pollution, it doesn't get rid of it. Diverting farm subsidies into the area of organic farming would be beneficial as would forcing government and agribusinesses to plant more diverse crops to maximize land usage.

"On a smaller scale you could eliminate, or at least reduce, the demand for certain products such as meats, dairy items, furs, and the like. Develop and use natural health and beauty products without using animals to experiment on. Speaking of which, there are also natural pet care products available. And, of course, promoting a vegetarian diet at the very least, if not a totally vegan diet."

"For the benefit of those who are not clear with respect to the difference between what constitutes a vegan versus a vegetarian diet would you care to explain?" I asked.

"Certainly," Mona replied. "A vegetarian diet means only elimi-
nating meat from the menu. It doesn't take into account other meat
products. Dairy products are usually consumed—milk, cheese,
and ice cream for example. One who commits to a vegan diet
refrains from eating anything that came from animals."

"You know that is going to be a tough sell for most of the population. You are talking about a huge change in dietary practices
and a complete change in philosophy as to what constitutes a
healthy, balanced diet. Not to mention the revision of whole
industries and their business practices."

"True, but when faced with the alternative."

"I'm just saying."

"I hear you, but that just means that you need to present the
benefits of such a change in a very positive manner."

"Suggestions, please," I said.

"Well, first of all, if people would switch to a vegetarian diet
that would mean an astronomical reduction in the slashing and
burning of the rainforests because the need for pasture land
would be greatly reduced. That would in turn eliminate the need
to chase indigenous peoples from their ancestral lands just so
your culture can gorge themselves on double burgers, filet
mignon, and, even worse, veal cutlets. You would also see a
drop in species extinction. Soil erosion and the resulting desertification of large areas of land would be much less of a problem
as the practice of over-farming would become unnecessary.
Fewer trees destroyed would mean a reduction of greenhouse
gases because of less carbon dioxide from the actual burning of
the forests and less methane as the result of a reduced livestock
population. Water pollution would be reduced because of less
solid waste from the livestock and water usage would drop sig-
nificantly. And, lastly, by partaking of a much healthier diet, you

would save a huge amount of money in the areas of both physical and emotional healthcare."

"Emotional healthcare? What does a vegetarian diet have to do with emotional health?" I asked.

"A great deal," Mona responded. "And, spiritual health as well."

"How so?"

"Have you ever seen the conditions at factory animal 'farms'?" Mona asked.

"Not in person, no," I answered. "Just what I have seen in photographs. That was bad enough."

"Then why do you ask me to explain? It should be self-explanatory."

"Indulge me then, will you please?"

"Very well. If the photos you saw depicted the scene accurately you saw animals, usually cattle, pigs, chickens, and turkeys living—and I use that term very loosely here—living in abominable conditions. They are kept crammed together in pens or cages so restrictive that they are unable to move, literally. They spend their entire lives this way, until comes time for their slaughter. Their lives consist of extreme physical abuse and terror. At the time of their death, they are literally consumed with fear, which has actual biological and physiological components. When humans eat these creatures, they consume this chemical fear. I propose that is one of the reasons that cultures such as yours are so violent and emotionally and psychologically ill. They are fear based."

"I have read about that," Lily said. "It is scary."

"And, it makes sense," I said. "With what we are learning about genetics and physiology the idea that we can ingest the fear of other species through our consumption of them is not that much of a stretch."

"Yeah, really," Lily said. "Think of all the ancient myths that have to do with warriors eating the heart or brain or whatever of their vanquished foe to take in their courage."

"It is not just a coincidence, Earthlings," Mona said. "Most of your myths are based on fact or actual events, even if in only a small way."

"I've heard that," I said.

"Oh, it's true. Now moving along, another financial component to this would be the huge amount of money saved on doctor, psychiatric, psychological, and hospital bills due to the various mental and emotional disorders brought on by your unhealthy eating habits, not to mention the resulting violence with its myriad financial consequences—the obvious medical ones along with the costs of the whole legal/penal system. Plus, as I said, the spiritual damage that you do to yourselves is enormous. At a subconscious, cellular level you know that what you are doing is immoral. Treating other creatures this way, especially when it is totally unnecessary for your survival, is a soul crime."

"I'll give you that," I said. "But, what about when it is necessary for survival?"

"When it is, it is. When in doubt, look to your ancestors for answers and look into your own heart. Your ancestors only took the life of another creature to provide for their essential needs to ensure survival: as a dietary supplement, for shelter and clothing, for tools and weapons. They didn't kill for sport or for trophies or just because they were in the mood for a little 'surf and turf' for dinner. They killed out of necessity and even then they treated the creature with respect and dignity and gratitude for giving up its life to provide for their needs.

"Stressing this as part of an educational program starting at a very young age would be of great benefit to future generations of humans and to all other life forms, including the planet. Also, you will want to get the meat and dairy industries out of your educational system and re-evaluate what is really a balanced diet and what is necessary for 'strong minds and bodies'."

"What are you saying?" Lily asked.

"As I alluded to earlier, these industries are disseminating false information. Their products are not necessary for a healthy diet. Most of the information that they espouse is pure propaganda to facilitate financial gain. The studies they cite to support their cause are undertaken for the most part by scientists on their payroll. Not exactly an unbiased group."

"I never thought of that," Lily said.

"Sensing a trend here?" Mona asked.

"Unfortunately."

"The bottom line is the less killing the better," Mona continued. "Whether you are talking about cattle and chickens, or wildflowers and trees, or coral reefs or rainforest...or humans. Each time extinction occurs there is one less strand holding the web of life together, therefore the web is weakened. When the extinction occur naturally, another specie adapts to fill the niche. But, the mass extinction's that have been taking place over the last few decades are not natural. When you save a species from extinction, whether it is the Spotted Owl or a simple fungus, you are potentially saving yourselves. It is widely known that there is interdependence between humans and the rest of nature, all of nature. What is not totally understood is how it all works. It would behoove you to err on the side of caution and prevent as much extinction as possible. Some of your scientists recognized this as far back as 1973 when they ratified the Endangered Species Act, for all the good it did. You never know what chain reaction of devastation you may prevent in the ecological framework by saving a specific species. You have to remember that there is no such thing as limitless growth in a closed ecosystem such as the planet you inhabit. You can either have negative interdependence such as dams and levees controlling flood waters, but in turn damaging the surrounding ecosystem, or positive interdependence in the manner of increasing organic farming in the Midwest, that leads to reduced leakage of nutrients into the Mississippi River, that leads to less stress on Caribbean coral reefs."

"Do you really think that we can pull it off?" I asked. "You are talking about changing the entire consciousness, not to mention the living habits, consuming habits, the very lifestyles and cultures of entire countries."

"Yes, I believe you can pull it off. If I didn't, I wouldn't be here. As you said to Dr. Don you humans are incredibly resourceful and creative and inventive and adaptable...and, intelligent. Also, compassionate and caring and, at your very core, loving. And, I am not just referring to those working for the environment or feeding the hungry or providing shelter for the homeless or trying to save the whales. I am talking about the presidents of big oil companies, tobacco barons, lobbyists, politicians, lawyers, and all the other folks who seem to be without souls. They, too, are capable of realizing who they really are and what is really important. They, too, are capable of seeing the light. They just need some help, someone to open their eyes. Don't alienate them. Use their talents. Incorporate them into the solution. They don't need to be destroyed, just re-directed. It is going to take an enormous amount of cooperation and teamwork to achieve the desired result. Integrate your scientists, religious leaders, environmental workers, Big Business, and government into one big problem solving entity."

"Oh, I can see that," Lily said. "Julia Butterfly Hill working with Charles Hurwitz. A match made in heaven."

"More than you know. She would be a perfect one to be involved in something like what I am proposing. Her ability to diffuse situations and humanize an event is extraordinary. Her ability to stay in a place of love, which is ultimately going to be the axis that the whole movement will rotate on, is essential. You bring these minds and spirits together and you will be able to evolve a new cultural perspective with a more workable system."

"I hope you are right about this," I said.

"I am Michael. I have seen it happen before."

"Really? Where?" Lily asked.

"No place that you have ever heard of. Trust me, it can be done. It will be done, on earth as it is in the heavens."

"Nice paraphrasing," I said.

"Thank you."

"How do you propose that we bring these ideological opposites together?" I asked.

"As to the activists, you will have to convince them that it is in their best interest, and the best interest of the planet, for them to be open to this type of problem solving. That shouldn't be too difficult. Most of them will welcome an open dialogue with their perceived enemies. Your challenge will in be bringing the establishment to the table. To do that, you will need to emphasize to Big Business and government that business practices that harm the environment and waste natural resources are less economical, and therefore less profitable, in the long run. Bottom line, when businesses and their shareholders are held accountable and forced to pay for the damage they cause, weather it be using up resources or polluting the environment, the result will be to force them to look into different ways of doing business. They will then be open to the fact that producing environmentally sound products and engaging in environmentally sound production is the profitable way to go. The activists can show them how to enact the necessary changes. Changes such as engaging in cradle-to-cradle practices."

"What is that?" I asked.

"Cradle-to-cradle refers to the practice of manufacturing wherein the product being produced can be reused and/or recycled into other products and everything that goes into making the product is environmentally friendly."

"That sounds like a great business strategy," Lily said.

"It is. The business eventually makes larger profits by not having to continually buy or harvest new resources for their production needs, the consumer benefits by getting better prices, and the environment wins by not getting decimated."

"It's a win-win-win trifecta."

"Yes. Using Eco-technology allows twice as much material welfare while using half the raw materials and producing half the waste and pollution. There are entire communities engaged in practices such as these. A few wonderful examples are the city of Kalundburg, Denmark and the town known as Gaviotas in Columbia.

"In Kalundburg, various industries such as the power plant, a pharmaceutical company, a biotechnology firm, a concrete plant, a fish farm, and a greenhouse are working together to produce an environmentally sound community through sharing technology and resources. In Gaviotas, which has been doing this since 1971 I might add, scientists, peasants, artists, ex-street kids, and Guahibo Indians among others have worked together to defy all odds for survival. This little community is 16 hours from the nearest major city, virtually in the middle of nowhere in a barren, rain-deprived savanna. Despite these disadvantages, not to mention the perennial political unrest of the country, they have established a community of unparalleled success for sustainable development. They have invented efficient wind turbines, solar collectors that work in the rain, soil-free systems to raise crops, ultra efficient pumps to tap deep aquifers that are so simple in design that they are operated by see-saws on the children's playground, and regenerated an ancient native rainforest. As I said, it can be done."

"As you know," I mentioned. "Lily is gaining expertise in the area of sustainable development."

"Of course," Mona acknowledged. "Would you care to expound a bit on this topic, Lily?"

"To the extent I can," Lily said. "Sustainable development encompasses living in material comfort and in peace with each other and all living things within the means of nature. Or, to put it in a more politically correct way, achieving human well being without exceeding the planet's ability to regenerate its resources and absorb waste."

"Sounds simple enough" I said. "And, I don't see how even the more conservative of people could find a problem with that philosophy."

"It is simple," Mona said. "And, the beauty of sustainable development is that, by necessity, it must remain flexible, adaptable, and creative to meet the needs of the different eco-systems where it is put into practice."

"That should appeal to my species," I said.

"Indeed it should. And, as you can see by the examples of Kalundburg and Gaviotas it is very doable. The key is getting people to commit." She paused here to give me a look. We both laughed at the commit thing, then she finished her thought, "To commit to all that sustainable development encompasses."

"Which is?" I asked.

"Which is a plan for each ecosystem that will include integrity, social justice, and financial well being not only in the present, but also for future generations. Sustainable development represents a way to sensibly and responsibly manage the use of resources, natural and otherwise, reduce waste and pollution, develop local economies, and improve the living conditions of disadvantaged populations. To do this, you will need to get people to accept that their survival is directly related to their interdependence not only with each other, but also with all of nature. Also, they will need to recognize that there are limits on natural resources and act accordingly. Next, they will need to practice fairness with all life when competing for resources. And, lastly, they will have to challenge the status quo."

"Gee, is that all?" I asked.

Mona ignored my sarcasm and said, "You could form communities where people could walk, bike, or use mass transit to take care of almost all of their daily activities: work, shopping, school, outdoor recreation, health care needs, and entertainment. The need for, and expensive of, automobiles would be reduced dramatically—not to mention the pollution factor."

"It does sound wonderful," Lily said.

"Yes, it does," I agreed.

"Earlier I touched on the topics of love," Mona said. I cringed. She acknowledged it with a nod and continued, "I know you think this will alienate some people."

"Some?" I said.

"Regardless," she replied. "The point is that you have been sent spiritual messengers throughout your history and you have chosen to ignore them for the most part or, at the very least, bastardize their message to suit your needs. Ironically, some of the writings that segments of your species hold as gospel address these very topics. But, as usual, you choose to ignore that part of the writings, as it would interfere with your 'pursuit of happiness'."

"What writings are you talking about?"

"Well, your Bible says: 'Ask now the beasts and they shall teach thee; and the fowls of the air and they shall teach thee; or, speak to the earth, and it shall teach thee; and the fishes of the waters shall declare unto thee.' And, 'For that which befalleth the sons of man befalleth the beast; even one thing befalleth them, as the one dieth, so dieth the other; yea, they have all one breath, so that a man hath no preeminence above a beast.' The Koran says: 'There is not an animal on earth, nor a flying creature on two wings, but they are peoples like unto you.' And, the Tao Te Ching says: 'Heaven, earth, and humans were created to be in harmony with one another, but humans lost the way and created disharmony'."

"Wow."

"But, there are other quotes," Lily said. "That people could cite that would argue the opposite."

"Like what?" I asked.

"Well, the Bible also says 'And God blessed them, and God said unto them, be fruitful and multiply, and replenish the Earth and subdue it; and have dominion over the fish of the sea, and over the fowl of the air, and over every living thing that moveth upon the Earth'."

"I'm sure," Mona said. "That people on each side of the issue can come up with quotes from various revered texts to bolster their views. I mean look at some of your adages and note the apparent contradictions: 'look before you leap' vs. 'he who hesitates is lost', 'birds of a feather flock together' vs. 'opposites attract'. The point is that these quotes, adages, and such should be taken in context. The key is to use common sense and follow your heart and your conscience. Do you honestly believe that

God set this whole thing in motion, watched it grow for all this time, then created man to destroy all of His beautiful and wondrous works? And, in just a few centuries, I might add?"

"No, that's not what I believe," I said. "But, there are those who will believe just that."

"True," Mona said. "There are some minds that will stay forever closed and totally impervious to change. But, you needn't change them all. Just enough to shift the consciousness and priorities of a minority of your culture. It takes but a few to inspire the many."

"You truly believe that, don't you?" I asked.

"Yes," Mona replied. "In an unstable system, a small ripple can have a great effect. You need look no further than the history of your country for a prime example."

"How so?" asked Lily.

"When your American Revolution was in it's infancy, less than one-third of the colonists supported the action. The rest either were apathetic to the cause or preferred to stay a part of the British Empire."

"You're kidding. I never knew that," I said.

"It's not something that is widely taught, but the point is that your founding fathers took a noble ideal—independence—and worked for it, fought for it, and, in some cases, died for it. The actions of a small core of idealists grew into the United States of America."

"I thought that you weren't too keen on the good old U.S. of A.?" I asked.

"I'm not keen at the way your country's politicians have sold out the people's interests to the corporations in their quest for financial riches and power. But, I am keen on the principles your country was founded upon: freedom for all, independence, liberty, equality, the pursuit of happiness...life.

The problem lies in the fact that these principles, along with the spiritual ones we spoke of earlier, have been compromised. But, with too many people on the planet for the return to tribal life to be an option, your democracy, when allowed to be of, by, and for the people, is still the best game in town. You just need to take it back from the special interests that presently control it. The ones who, left uncontested, will continue on their present course of environmental destruction. It's time for another revolution."

"I agree," I said.

"Good. Then may I suggest a few more areas that need addressing?"

"By all means, continue."

"I believe that, along with the options that I've already mapped out, the arrogant belief that God favors one country or one culture or one race or one religion over another must be abandoned. As does the mistaken notion that military 'might' is synonymous with 'right'. Bottom line, your job is to give the people something new to believe in and to lead by example." She stopped at this point and took a deep breath. I noticed a look of melancholy sweep across her face.

"I get the sense that we are reaching the end of our collaboration," I said.

"Yes, we are. My crew and I have work to do elsewhere."

"Will you be back? I mean, in this lifetime?" I asked.

"Of course, Michael. If nothing else, to gauge the level of your 'commitment'," she said while doing the quotation mark thing with her hands and giving me her most brilliant smile.

"That's cute, Mona," I said, returning her smile.

"But, before we adjourn Spats (the Spaceman), Dr. Don, and Demi would like to meet Lily and say their good byes to you."

"Oh, boy," I said. *Just when I thought I might get off easy.*

"Cool," Lily said. She looked over her shoulder at the door of the auditorium.

"I don't think that's where they will be coming from," I said.

"What do you mean by that?" Lily asked.

"I mean that I think you are going to get to witness tunneling," I answered, looking to Mona for confirmation. Mona nodded 'yes' and looked toward the opposite end of the table. Lily and I followed her gaze…and there stood Spats (the Spaceman)…then Dr. Don…then Demi—looking especially lovely, I might add. *Damn it.*

"Wow," Lily said. *I hope she is referring to the tunneling and not to Demi?*

"Oh, boy," I said.

"That is way cool," Lily said. *So far, so good.*

Mona stood up to greet them as did Lily and I. Dr. Don and Demi approached us while Spats (the Spaceman) hung back. I think three beautiful women at one time were sensory overload for him. We exchanged introductions and greetings with Dr. Don and Demi and eventually coaxed Spats (the Spaceman) to join us. Thankfully, Demi was able to win over Lily much as Mona had, thereby greatly reducing the tension in the room…especially mine. And, to her credit, Lily handled the 'Demi situation' with grace—although I did get a look that I read as meaning we would be discussing it when she got me alone.

We ended up talking for hours about the book, tunneling, intuiting, and a variety of other subjects. The crew gave us a few more suggestions and gave us their enthusiastic encouragement. As things began to wind down, Lily drew Mona to the side and exchanged a few words with her. Mona nodded and smiled, then stepped a few paces away from Lily. I approached Lily and inquired as to what was happening. She said she had expressed an interest to meet Lean Gene.

"And?" I asked.

"She is discussing it with him."

"Oh, boy," I said.

"Is that your new mantra or what?" Lily said.

About then Lean Gene tunneled onto the scene. I don't know what I was expecting (probably a World Wrestling Federation look-a-like based on my experience with his voice and demeanor) but, it wasn't what I got. He turned out to be the perfect counterpart to Mona. Same age and hard to pin down appearance. Where Mona's looks changed in respect to ethnicity, Lean Gene's appearance morphed through a gamut of animal images: from one angle he looked feline: sleek, graceful, exotic; from another he looked more avian: long nose, piercing eyes; from yet another perspective he took on the traits of the equine: robust, powerful, noble. It was disconcerting, but strangely beautiful.

After hugging Mona and briefly conferring with her, he exchanged greetings with the other crewmembers. At last, Mona made our introductions. He acknowledged Lily with a kiss on each cheek, European style, then turned to me. "Michael, it is a pleasure to finally meet you," he said grasping my offered hand between both of his. His touch and manner exuded warmth and compassion. He was as disarming and genuine as his wife was. I admit that I was taken aback and instantly won over. It didn't hurt that his voice was a few octaves higher in person than the one that came through Mona.

"Pleased to meet you, too, Lean Gene," I said.

"Just Gene, please."

"Okay, Gene."

"We very much appreciate your efforts, Michael."

"Thank you. But, you and Mona and your crewmembers have done all the work. I am just recording and reporting it."

"I wish it were that simple. But, make no mistake about it, Michael, what you have undertaken here is not going to be easy nor is it always going to be pleasant."

"Nothing ever is," I responded.

Mona and Lily made their way over to where Lean Gene and I were standing. The crewmembers stayed back and talked amongst themselves.

"It's going to be like the dieting experience you related earlier," Mona said. "You are going to have to muster all of your resources at times just to carry on. And, you can't be looking too far ahead or viewing this as a financial venture, either."

"I know all of that, Mona. But, if I do happen to make a bit of money along the way I won't complain, nor will I feel guilty about it either."

"Me neither," Lily agreed. "But, I understand what you are saying Mona. It mustn't be the focus or the motivating factor. That's one of the main factors in how we got in this predicament in the first place."

"Exactly," Mona said. "There is nothing wrong with monetary gain, per se. I'm just saying that the real motivation must come from within. A desire to engage in the activity for the simple reason that it is the right thing to do."

"I believe," Lean Gene said. "That the term you are looking for is 'altruism'."

"Yes, altruism," Mona said. She gave Lean Gene her most brilliant smile. "Doing it purely for the inner reward."

"Believe it or not, I understand that concept," I said. "As does Lily, I'm sure."

"Yes. I have experienced the wonderful feelings of giving for the pure joy of giving," Lily said.

"I always experience it," I said. "As an overwhelming sense of gratitude, that I am exactly where I am supposed to be and doing exactly what I am supposed to be doing. No other feeling like it on Earth."

"Or, anywhere else for that matter," Mona said.

"The point is," Lean Gene said. "That you have to keep the big picture in mind at all times. Any residuals that flow from that will be icing on the cake."

"I get it, already," I said.

"Yes," Mona said. "No need to belabor the point. We wouldn't have chosen you if we didn't think you could handle the responsibility."

"This is true," Lean Gene agreed.

"Well, thank you," I said.

The other crewmembers drifted over to bid Lily and I adieu. There was an awkward moment when Demi and I said our good byes. We did a semi-vaudeville act of alternately offering to shake hands or hug, finally settling on the double-cheek buss that Lean Gene had demonstrated on Lily. Spats (the Spaceman), Dr. Don, and Demi then tunneled away after promising to visit again the next time their schedule allowed.

Mona, Lean Gene, Lily, and I talked for another hour or so, reluctant to bring our time together to a close. But, as the saying goes 'all good things must end' and they made ready for their departure. Lean Gene and I complimented each other on our choice of female companionship and said our farewells as Mona and Lily did the same (at least as far as saying farewells at any rate). Mona turned to me and Lily took her cue, taking Lean Gene by the arm and walking away a few paces.

"I'm certainly going to miss you, Mona," I said.

"I know, dear. As I will you." And, intuiting my thoughts yet again she added, "And, no I don't know exactly when we will return."

"That I won't miss," I said.

"I suspect not," Mona said, beaming. "But wait until you have mastered it. Then we will see how you feel about it."

"If I master it," I corrected her.

"You will, Michael. Believe me."

"Yes ma'am."

"I know you are worried about the responsibility that you have agreed to take on, but your worries are unfounded."

"Intellectually I know that. Now, I just need to believe it at gut level."

"That will come, too. Just remember to stay humble, Michael. Down to earth."

Lean Gene called to her. "It's time, Light."

"Be right there, dear," she replied.

"Light?" I inquired.

"Don't ask." We walked over to where Lean Gene stood waiting with Lily. Mona opened her arms to me in invitation and we hugged fiercely for a long moment. As she stepped away from me, she whispered in my ear, "Good luck with Lily. I hope you two can work things out. You are deserving of each other." I'm still not certain how she meant that last part, but in deference to Lily, I prefer to give her the benefit of the doubt.

She took Lean Gene's hand. I nodded to them and smiled. Then they were gone.

Epilogue

Well, that's my story (and I'm sticking to it) of my experience with the entity our culture refers to as Mother Nature. After many trials and tribulations that I won't bore you with, the book was published (yes, I know, another comment on the obvious). My hope now is that you have come to believe that the concerns raised herein are real and valid and that you want to become a part of the solution.

If you have reached that conclusion, then I suggest you start by reading any one of the wonderful books or magazines or visiting one of the Web Sites listed in the References section. All of those resources will also have references of their own and ideas on how to proceed. The list is growing daily. If the immensity of the problems and suggested solutions that you read about seem overwhelming, maybe you can start with something small: recycling your paper and plastic products; buying only environmentally friendly products; adopting an area of your town for litter patrol; car pooling or using mass transit. You will not only be helping the environment; you will also be setting an example for your families, your friends, your co-workers...and, your children and grandchildren.

If you are thinking 'I'm only one person. Can I really make a difference?' I have but three words for you: Bush versus Gore. Please, make your voice heard. Add it to the chorus of people just like you who are already involved in this cause. I firmly believe that we *can* make a difference, that we *can* heal the planet. If you are among those who are still skeptical or who just don't want to get involved...please, return to page one and reread.

As for Lily and I, we are still trying to 'work it out' as Mona put it. What will result from our efforts? I don't know. There are no guarantees. But, there is always hope. So, we are still plugging away, giving it our best shot, and that's all (Mona and) I ask of you.

Finally, to all of you, on behalf of Mona, Lean Gene, and their crewmembers; all of the other good folks already taking action; and, Lily and myself (ouch, and Nikita), thank you for taking the time to read this book. Carpe Diem.

Sincerely, Michael E. Rice 04/2001

About the Author

Michael E. Rice has written two screen-plays and numerous short stories. He works as a bookseller in addition to lectur-ing at a DUI program. He possesses a BA in Psychology and an MS in Administration. He resides in Dayton, Ohio and is at work on his second book. He can be reached at: Mikerophone1@cs.com and at his Web site: www.Downtoearthbook.com.

References

BOOKS

50 Simple Things You Can Do To Save the Earth by The Earth Works Group

A Brief History of Time by Hawking, Stephen

A Fish Caught in Time by Weinberg, Samantha

A Pocket Guide to Environmental Bad Guys by Ridgeway, James & St. Clair, Jeffrey

A Reason for Hope by Goodall, Jane

A Reason to Vote by Roth, Robert

A Sand County Almanac by Leopold, Aldo

A Time Before History by Tudge, Colin

Act Now, Apologize Later by Werbach, Adam

Amazon Stranger by Tidwell, Mike

And the Waters Turned to Blood by Barker, Rodney

Believing Cassandra by AtKisson, Alan

Biophilia & Consilience by Wilson, Edward O.

Bound to the Earth by Swan, James A. & Roberta

But Not a Drop to Drink by Coffel, Steve

Callings by Levoy, Gregg

Celestine Prophecy, The Tenth Insight, Secrets of Shambala, & Celestine Vision by Redfield, James

Conscious Evolution by Hubbard, Barbara Marx

Contemporary World Issues by Miller, Ruby & Miller, E. Willard

Conversations With God: Books 1,2, & 3; Friendship With God; & Communion With God by Walsch, Neale Donald

Culture Jam by Lasn, Kalle

Cybergrace by Cobb, Jennifer

Darkness in El Dorado by Tierney, Patrick

Diet for a New America by Robbins, John

Driving Mr. Albert by Paterniti, Michael

Earth For Sale by Tokar, Brian

Earth in the Balance by Gore, Al

Ecology of Commerce & Natural Capitalism by Hawken, Paul

Einstein's Brainchild by Parker, Barry

Finding Darwin's God by Miller, Kenneth R.

Galileo's Daughter by Sobel, Dava

Gaviotas by Weisman, Alan

Genome by Ridley, Matt

Global Warming by Bender, David & Leone, Bruno

God and the New Physics & The Mind of God by Davies, Paul

Going Local by Shuman, Michael H.

Guns, Germs, And Steel by Diamond, Jared

Into Thin Air by Kidd, J.S. & Renee

Ishmael, The Story of B, My Ishmael, Providence, & Beyond Civilization by Quinn, Daniel

Last Hours of Ancient Sunlight & The Prophets Way by Hartmann, Thom

Laws of the Spirit by Millman, Dan

Lie My Teacher Told Me by Loewen, James W.

Life: The Movie by Gabler, Neal

Living Downstream by Steingraber, Sandra

Mutant Message Down Under & Message From Forever by Morgan, Marlo

Nature's Keepers by Tobias, Michael

Not In My Back Yard: The Handbook by Morris, J.A.

Our Poisoned Waters by Dolan, Edward F.

Our Stolen Future by Colbern, Theo, Dumandski, Dianne, & Myers, John Peterson

Ozone by Gay, Kathlyn

Rainforests For Beginners by Rosenblatt, Naomi

Reflections of Eden by Galdikas, Birute

Reworking Success by Theobald, Robert

River Out of Eden by Dawkins, Richard

Silent Spring by Carson, Rachel

Spirit in the Genes by Morrison, Reg

State of the World 2000 by Worldwatch Institute

Stuff: The Secret Lives of Everyday Things by Ryan, John C. & Durning, Alan

Teaching a Stone to Talk by Dillard, Annie

The Cathedral Within by Shore, Bill

The Coming Global Superstorm by Bell, Art & Strieber, Whitley

The Elegant Universe by Greene, Brian

The Fifth Miracle by Davies, Paul

The Forgiving Air by Somerville, Richard C.J.

The Future of Love by Kingma, Daphne Rose

The Healing of America by Williamson, Marianne

The Legacy of Luna by Hill, Julia Butterfly

The Monkey Wrench Gang, Hayduke Lives, & Desert Solitaire by Abbey, Edward

The Politics of Meaning by Lerner, Michael

The Science of God by Schroeder, Gerald C.

The Sixth Extinction by Leakey, Richard E. & Lewin, Roger

The Twilight of American Culture

The Wasted Oceans by Bulloch, David K.

To the Elephant Graveyard by Hall, Tarquin

Watching From the Edge of Extinction by Sterns, Beverly P. & Sterns, S.C.

When Corporations Rule the World by Korten, David C.

MAGAZINES

Business Ethics Magazine

EarthWatch Institute

Greenpeace

Rachel's Environmental & Health Weekly

Sierra

Solar Today

Utne Reader

Whole Earth Catalog

WEB SITES

Adbusters.org

ALCnet.org

BeyondCreation.org

Chemsafety.gov

CleanAir.gov

CleanWater.gov

Corpwatch.org

Downtoearthbook.com

Earthsystems.org

Earthtimes.org

Ecomail.com

Ecotopia.org

ENN.com

ENS.lycos.com

Env.about.com

Envirolink.netforchange.com

Enviroweb.org

EPA.gov

EPALAW.com

FOE.org

Freezone.co.uk

LTA.org

Motherjones.com

Natural-law.org

Public-i.org
Rainforest.Care2.com
RAN.org
Raysweb.net
Recreating.org
Renaissancealliance.org
SavetheRedwoods.org
Sierraclub.org
Soyouwanna.com
TNC.org
Vegan.com
Veganoutreach.org
Wetlands.com